COMPOSITE MATERIALS ENGINEERING
ENGINEERING

Modeling and Technology

COMPOSITE MATERIALS ENGINEERING

Modeling and Technology

Edited by
Alexander V. Vakhrushev, DSc
A. K. Haghi, PhD

APPLE ACADEMIC PRESS

Apple Academic Press Inc.
4164 Lakeshore Road
Burlington ON L7L 1A4
Canada

Apple Academic Press Inc.
1265 Goldenrod Circle NE
Palm Bay, Florida 32905
USA

© 2020 by Apple Academic Press, Inc.

First issued in paperback 2021

Exclusive worldwide distribution by CRC Press, a member of Taylor & Francis Group

No claim to original U.S. Government works

ISBN 13: 978-1-77463-472-1 (pbk)
ISBN 13: 978-1-77188-796-0 (hbk)

Library and Archives Canada Cataloguing in Publication

Title: Composite materials engineering : modeling and technology / edited by Alexander V. Vakhrushev, DSc, A.K. Haghi, PhD.

Names: Vakhrushev, Alexander V., editor. | Haghi, A. K., editor.

Description: Includes bibliographical references and index.

Identifiers: Canadiana (print) 2019016574X | Canadiana (ebook) 20190165774 | ISBN 9781771887960 (hardcover) | ISBN 9780429242762 (ebook)

Subjects: LCSH: Composite materials.

Classification: LCC TA418.9.C6 C623 2020 | DDC 620.1/18—dc23

Library of Congress Cataloging-in-Publication Data

Names: Vakhrushev, Alexander V., editor. | Haghi, A. K., editor.

Title: Composite materials engineering : modeling and technology / edited by Alexander V. Vakhrushev, DSc, A. K. Haghi, PhD.

Other titles: Composite materials engineering (Apple Academic Press)

Description: 1st edition. | Oakville, ON ; Palm Bay, Florida : Apple Academic Press, Inc., 2019. | Includes bibliographical references and index. | Summary: "This book provides a compilation of innovative fabrication strategies and utilization methodologies that are frequently adopted in the advanced composite materials community. It addresses developing appropriate composites to efficiently utilize macro- and nanoscale features. It covers a selection of key aspects of composite materials, including history, reinforcements, matrix materials, mechanical properties, physical properties, theory, and applications. The volume reviews the research developments of a number of widely studied composite materials with different matrices key features. Contains new coverage of nanocomposites that reflects the latest theoretical and engineering and industrial applications of composite materials. Provides design methods with numerical information and technical formulations needed for researchers. Presents a critical review of progress in research and development on composite materials. Offers comments on future research direction and ideas for product development"-- Provided by publisher.

Identifiers: LCCN 2019035117 (print) | LCCN 2019035118 (ebook) | ISBN 9781771887960 (hardcover) | ISBN 9780429242762 (ebook)

Subjects: LCSH: Composite materials.

Classification: LCC TA418.9.C6 C569 2019 (print) | LCC TA418.9.C6 (ebook) | DDC 620.1/18--dc23

LC record available at https://lccn.loc.gov/2019035117

LC ebook record available at https://lccn.loc.gov/2019035118

Apple Academic Press also publishes its books in a variety of electronic formats. Some content that appears in print may not be available in electronic format. For information about Apple Academic Press products, visit our website at **www.appleacademicpress.com** and the CRC Press website at **www.crcpress.com**

About the Editors

Alexander V. Vakhrushev, DSc

Professor, M.T. Kalashnikov Izhevsk State Technical University, Izhevsk, Russia; Head, Department of Nanotechnology and Microsystems of Kalashnikov Izhevsk State Technical University, Russia

Alexander V. Vakhrushev, DSc, is a professor at the M.T. Kalashnikov Izhevsk State Technical University in Izhevsk, Russia, where he teaches theory, calculating, and design of nano- and microsystems. He is also the Chief Researcher of the Department of Information-Measuring Systems of the Institute of Mechanics of the Ural Branch of the Russian Academy of Sciences and Head of the Department of Nanotechnology and Microsystems of Kalashnikov Izhevsk State Technical University. He is a Corresponding Member of the Russian Engineering Academy. He has over 400 publications to his name, including monographs, articles, reports, reviews, and patents. He has received several awards, including an Academician A. F. Sidorov Prize from the Ural Division of the Russian Academy of Sciences for significant contribution to the creation of the theoretical fundamentals of physical processes taking place in multi-level nanosystems and Honorable Scientist of the Udmurt Republic. He is currently a member of the editorial board of several journals, including *Computational Continuum Mechanics, Chemical Physics and Mesoscopia*, and *Nanobuild*. His research interests include multiscale mathematical modeling of physical-chemical processes into the nano-hetero systems at nano-, micro-, and macro-levels; static and dynamic interaction of nanoelements; and basic laws relating the structure and macro characteristics of nano-hetero structures.

A. K. Haghi, PhD

Former Editor-in-Chief, International Journal of Chemoinformatics and Chemical Engineering and Polymers Research Journal;
Member, Canadian Research and Development Center of Sciences and Cultures (CRDCSC), Montreal, Quebec, Canada,
E-mail: AKHaghi@Yahoo.com

A. K. Haghi, PhD, is the author and editor of 165 books, as well as 1000 published papers in various journals and conference proceedings. Dr. Haghi has received several grants, consulted for a number of major corporations, and is a frequent speaker to national and international audiences. Since 1983, he served as a professor at several universities. He is the former Editor-in-Chief of the *International Journal of Chemoinformatics and Chemical Engineering* and *Polymers Research Journal* and on the editorial boards of many international journals. He is also a member of the Canadian Research and Development Center of Sciences and Cultures (CRDCSC), Montreal, Quebec, Canada. He holds a BSc in urban and environmental engineering from the University of North Carolina (USA), an MSc in mechanical engineering from North Carolina A&T State University (USA), a DEA in applied mechanics, acoustics, and materials from the Université de Technologie de Compiègne (France), and a PhD in engineering sciences from Université de Franche-Comté (France).

Contents

Contributors

M. Avaliani
Iv. Javakhishvili Tbilisi State University, R. Agladze Institute of Inorganic Chemistry and Electrochemistry, 0186 Mindeli Str., 11, Tbilisi, Georgia,
E-mails: avaliani21@hotmail.com, marine.avaliani@tsu.ge

N. Barnovi
Iv. Javakhishvili Tbilisi State University, R. Agladze Institute of Inorganic Chemistry and Electrochemistry, 0186 Mindeli Str., 11, Tbilisi, Georgia

Gloria Castellano
Departamento de Ciencias Experimentales y Matemáticas, Facultad de Veterinaria y Ciencias Experimentales, Universidad Católica de Valencia San Vicente Mártir, Guillem de Castro-94, E-46001 València, Spain

V. Chagelishvili
Iv. Javakhishvili Tbilisi State University, R. Agladze Institute of Inorganic Chemistry and Electrochemistry, 0186 Mindeli Str., 11, Tbilisi, Georgia

Tanmoy Chakraborty
Department of Chemistry; Manipal University Jaipur, DehmiKalan, Jaipur – 303007, India, E-mails: tanmoychem@gmail.com; tanmoy.chakraborty@jaipur.manipal.edu

N. Esakia
Iv. Javakhishvili Tbilisi State University, Faculty of Exact and Natural Sciences, Department of Chemistry, 0179 Chavchavadze ave. 3, Tbilisi, Georgia

V. B. Golubchikov
Scientific and Production Company "Nord," Perm, Russia

A. V. Gumovskii
Department of Nanotechnology and Microsystems, Technic Kalashnikov Izhevsk State Technical University, Izhevsk, Russia, E-mail: gumma.andres@gmail.com

M. Gvelesiani
Iv. Javakhishvili Tbilisi State University, R. Agladze Institute of Inorganic Chemistry and Electrochemistry, 0186 Mindeli Str., 11, Tbilisi, Georgia

Shrikaant Kulkarni
Department of Chemical Engineering, Vishwakarma Institute of Technology, Pune (M.S.), India, E-mail: shrikaant.kulkarni@vit.edu

Ajay Kumar
Department of Mechatronics Engineering, Manipal University Jaipur, DehmiKalan, Jaipur – 303007, India

Sh. Makhatadze
Iv. Javakhishvili Tbilisi State University, R. Agladze Institute of Inorganic Chemistry and Electrochemistry, 0186 Mindeli Str., 11, Tbilisi, Georgia

Sukanchan Palit
43, Judges Bagan, Post-Office-Haridevpur, Kolkata – 700082, India,
Tel.: 0091-8958728093, E-mails: sukanchan68@gmail.com, sukanchan92@gmail.com

Prabhat Ranjan
Department of Mechatronics Engineering, Manipal University Jaipur, DehmiKalan,
Jaipur – 303007, India

Ana C. F. Ribeiro
Centro de Química, Department of Chemistry, University of Coimbra, 3004-535 Coimbra,
Portugal, E-mail: anacfrib@ci.uc.pt

A. A. Shushkov
Udmurt Federal Research Center of the Ural Branch of the Russian Academy of Sciences,
Institute of Mechanics, Department of Mechanics of Nanostructures, Izhevsk, Russia

D. Starokadomsky
Department of Composites, Chuiko Institute of Surface Chemistry, NAS of Ukraine, 03164, Kiev,
Ukraine, E-mail: km80@ukr.net

Francisco Torrens
Institut Universitari de Ciència Molecular, Universitat de València, Edifici d'Instituts de Paterna,
P. O. Box 22085, E-46071 València, Spain, E-mail: torrens@uv.es

A. V. Vakhrushev
Department of Mechanics of Nanostructures, Institute of Mechanics,
Udmurt Federal Research Center, Ural Division, Russian Academy of Sciences,
Izhevsk, Russia | Department of Nanotechnology and Microsystems,
Technic Kalashnikov Izhevsk State Technical University, Izhevsk, Russia,
E-mail: vakhrushev-a@yandex.ru

A. V. Zhivotkov
Scientific and Production Company "Nord," Perm, Russia, E-mail: nord59r@mail.ru

Abbreviations

AAS	atomic absorption spectrometry
AFM	atomic force microscopy
B3LYP	Becke's three-parameter Lee-Yang-Parr
BFC	biofuel elements
CA	catalytic activity
CDFT	conceptual density functional theory
CNT	c-nanotube
CVD	chemical vapor deposition
CVDG	chemical vapor deposition graphene
DFT	density functional theory
DMTA	dynamic mechanical thermal analysis
DRI	differential refractive index
DRS	dielectric relaxation spectroscopy
DSC	differential scanning calorimetry
DVM	digital voltmeter
ECD	electrocodeposition
EPGG	epitaxial growth graphene
ESR	electron spin resonance
FEM	finite element method
FLG	few-layer GR
FTIR	Fourier transform infrared spectroscopy
GC	gas chromatography
GIC	intercalated graphite compound
GICG	graphene intercalated graphite compounds
GMR	giant magnetic random
GO	graphene oxide
GOx	glucose oxidase
GPC	gel permeation chromatography
GRs	graphenes
HETP	height equivalent to a theoretical plate
HOMO	highest occupied molecular orbital
HPLC	high-performance liquid chromatography
IC	ion chromatography

ICP	inductively coupled plasma
IEC	ion-exchange chromatography
LALLS	low-angle laser light scattering
LC	liquid chromatography
LLC	liquid/liquid chromatography
LSC	liquid/solid chromatography
LUMO	lowest unoccupied molecular orbital
MMEC	metal matrix composite electrochemical coatings
MRAM	magnetic random access memory
MS	mass spectrometry
NCs	nanocomposites
NMR	nuclear magnetic resonance
NMs	nanomaterials
NP	nanoparticle
NS	nanoscience
NT	nanotechnology
OM	optical microscopy
PMMA	polymethylmethacrylate
PTE	periodic table of the elements
QDs	quantum dots
QML	quantum machine learning
RS	Raman spectroscopy
RT	room temperature
SEC	size exclusion chromatography
SEM	scanning electron microscopy
SMMs	single-molecule magnets
SWCNT	single-wall c-nanotube
TA	thermal analysis
TCPP	tetrakis(4-carboxyphenyl)porphyrin
TEM	transmission electron microscopy
TGA	thermogravimetric analysis
TLC	thin-layer chromatography
TMA	thermomechanical analysis
TMR	tunneling magnetic random
VDW	Van Der Waals
VIS	viscometer
WAXRD	wide-angle x-ray diffraction
XPS	x-ray photoelectron spectroscopy
XRD	x-ray diffraction

Preface

Composites materials are basically the combining of unique properties of materials to have synergistic effects. A combination of materials is needed to adapt to certain properties for any application area. Composite materials have grown rapidly in their applications; they will no doubt continue to do so. Composite materials are versatile materials with a variety of applications. A greater use of composite materials in many areas of engineering has led to a greater demand for engineers versed in the design of structures made from such materials.

Composites are now established engineering materials, offering weight savings and longevity to a rapidly expanding range of applications. This book is unique in that and it not only offers a current analysis of mechanics and properties, but also examines the latest advances in theory and applications and design aspects involving composites. This book also provides the foundation for traditional basic composite material mechanics, making it a comprehensive reference on this topic.

Key features of this book:
Each chapter begins with an overview of the key points to be addressed:

- Contains new coverage of nanocomposites.
- Reflects the latest theoretical and engineering applications.
- Provides design methods with numerical information and technical formulations needed for researchers.
- Systematic and comprehensive information on composite materials.
- Critical review of progress in research and development on composite materials.
- Comment on future research direction and ideas for product development.
- Presents advanced material on composite structures.
- Provides industrial applications of composite materials.

CHAPTER 1

Conformational Analysis of Ag-Au Nanoalloy Clusters: A CDFT Approach

PRABHAT RANJAN,[1] AJAY KUMAR,[1] and TANMOY CHAKRABORTY[2]

[1]*Department of Mechatronics Engineering, Manipal University Jaipur, DehmiKalan, Jaipur – 303007, India*

[2]*Department of Chemistry, Manipal University Jaipur, DehmiKalan, Jaipur – 303007, India, E-mails: tanmoychem@gmail.com; tanmoy.chakraborty@jaipur.manipal.edu*

ABSTRACT

The study of Ag-Au nanoalloy clusters is of considerable interest due to a unique optical and electronic and large range of technological applications in biosciences, material sciences, and catalysis. Density functional theory (DFT) is the most efficient technique of quantum mechanics to explore the electronic properties of matter. Conceptual density functional theory (CDFT) has been proven to be an essential tool to correlate the experimental properties of clusters with CDFT-based descriptors. In this chapter, conformational analysis of Ag-Au clusters has been studied by using DFT methodology. We have computed CDFT-based descriptors viz. HOMO-LUMO energy gap, electronegativity, hardness, softness, electrophilicity index, and dipole moment of trimers by changing the angle between the atoms.

1.1 INTRODUCTION

It is a well-known fact that structure and property are very much closely related to each other [1]. There is not any particular method to determine the symmetry structure of a molecule. Though, density functional theory

(DFT)-based global descriptors like HOMO-LUMO energy gap, hardness, softness, electronegativity, and electrophilicity index have been invoked to relate structure and property of molecules [1–3]. Even if there is not any distinctive relationship between structure and properties, the structure and property of molecules are interrelated. Nowadays, science and technology has been developed to a new level for understanding the biological and molecular activities of the system [4–7]. Researchers are examining the descriptors in terms of mathematical values that pronounce the structure and physicochemical properties of the molecular system [8–11]. But, for the prediction of the relationship between structure and property, it is very important to investigate the electronic structure and operational forces on the molecular system [10, 11]. Though some researchers have tried to show the relationship between properties and reactivity of a molecule with reference to its equilibrium structure, a molecule rarely stays in its stationary equilibrium shape. The physical process of the dynamics of internal alternation begins the isomerization process mechanism, and it produces countless conformations [13, 14]. The conformational analysis has an acute result on biological activities and response on the result of various stereochemical reactions [1]. Therefore, an insight of the relative energies and the mechanism of the advancement of conformational analysis will be able to predict information regarding the sensitivity, the spatial arrangement of atoms and their effects on the physicochemical properties and product dissemination in reaction [15]. Thus, the study of conformations as to the source of a barrier to internal alternation within a molecule is of high importance to theoretical, experimental, and biological scientists.

1.2   COMPUTATIONAL DETAILS

Since the last couple of years, DFT has been a dominant and effective computational technique for bimetallic and multi-metallic clusters. DFT methods are open to many new innovative fields in material science, physics, chemistry, surface science, nanotechnology, biology, and earth sciences [16]. Among all the DFT approximations, the hybrid functional Becke's three-parameter Lee-Yang-Parr exchange-correlation functional (B3LYP) has been proven very efficient and used successfully for bimetallic clusters [17–20]. In this section, conformational analysis of bimetallic nanoalloy clusters has been performed with the help of hybrid functional Becke's three-parameter Lee-Yang-Parr exchange-correlation functional (B3LYP). The basis set LanL2dz has high accuracy for metallic clusters, which has been recently analyzed by researchers [17, 21, 22]. For optimization purpose, Becke's three-parameter Lee-Yang-Parr (B3LYP) exchange-correlation with basis set LanL2dz has been adopted. All the modeling and structural optimization of compounds have been performed using the Gaussian 03 software package [23] within the DFT framework.

Invoking Koopmans' approximation [24, 25], we have calculated ionization energy (I) and electron affinity (A) of all the nanoalloys using the following ansatz:

$$I = -\varepsilon_{HOMO} \tag{1}$$

$$A = -\varepsilon_{LUMO} \tag{2}$$

Thereafter, using I and A, the conceptual DFT-based descriptors viz. electronegativity (χ), global hardness (η), molecular softness (S) and electrophilicity index (ω) have been computed. The equations used for such calculations are as follows:

$$\chi = -\mu = \frac{I + A}{2} \tag{3}$$

where μ represents the chemical potential of the system.

$$\eta = \frac{I - A}{2} \tag{4}$$

$$S = \frac{1}{2\eta} \tag{5}$$

$$\omega = \frac{\mu^2}{2\eta} \tag{6}$$

1.3   RESULTS AND DISCUSSION

In this study, conformational analysis of three atoms of bimetallic clusters AgAu, i.e., $AgAu_2$, and Ag_2Au have been performed. The orbital energies in the form of highest occupied molecular orbital (HOMO)–lowest unoccupied molecular orbital (LUMO) energy gap along with DFT-based descriptors have been investigated as a function of dihedral angles.

1.3.1   AGAU$_2$ NANOALLOY CLUSTERS

a.   In Figure 1.1, Au atom is fixed at the center, and other species, i.e., Ag and Au placed at corners are not fixed.

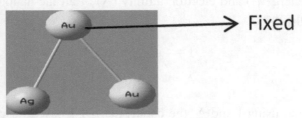

FIGURE 1.1   Cluster AgAu$_2$ with Au atom fixed at the center.

The orbital energies in the form of HOMO-LUMO gap along with computed DFT-based descriptors viz. electronegativity, hardness, softness, electrophilicity index, and dipole moment have been analyzed as a function of dihedral angles and reported in Table 1.1. Figure 1.1 and Table 1.1 establish that the preferred conformation of the instant cluster has a linear structure with Au atom fixed at the center. The maximum HOMO-LUMO energy gap is obtained at the dihedral angle of 180°, while the least gap is obtained at a dihedral angle of 20°. The instant cluster has constant HOMO-LUMO gap, and DFT-based descriptors for dihedral angles varying between 20°–80° and 90°–170°; however, an exception has been marked at an angle 60°.

b.   In Figure 1.2, the triangular structure of AgAu$_2$ cluster is formed with fixed Au atom at the center, and other Au and Ag atoms are placed at the corners of the triangle. The computed physicochemical properties of all the generated conformations are shown in Table 1.2 with the corresponding dihedral angles.

TABLE 1.1 Computed DFT-Based Descriptors of AgAu$_2$ Cluster with Au Atom Fixed at the Center

Angles (in Degrees)	HOMO-LUMO Gap (eV)	Electro-negativity (eV)	Hardness (eV)	Softness (eV)	Electro-philicity Index (eV)	Dipole Moment (Debye)	Symmetry
20	1.605	4.258	0.803	0.623	11.295	3.223	C$_s$
30	1.605	4.258	0.803	0.623	11.295	3.222	C$_s$
40	1.605	4.258	0.803	0.623	11.295	3.222	C$_s$
50	1.605	4.258	0.803	0.623	11.295	3.226	C$_s$
60	1.687	4.218	0.844	0.593	10.544	3.752	C$_s$
70	1.605	4.258	0.803	0.623	11.295	3.225	C$_s$
80	1.605	4.258	0.803	0.623	11.295	3.225	C$_s$
90	2.694	4.394	1.347	0.371	7.169	3.341	C$_s$
100	2.694	4.394	1.347	0.371	7.169	3.341	C$_s$
110	2.694	4.394	1.347	0.371	7.169	3.341	C$_s$
120	2.694	4.394	1.347	0.371	7.169	3.341	C$_s$
130	2.694	4.394	1.347	0.371	7.169	3.34	C$_s$
140	2.694	4.394	1.347	0.371	7.169	3.34	C$_s$
150	2.694	4.394	1.347	0.371	7.169	3.341	C$_s$
160	2.694	4.394	1.347	0.371	7.169	3.342	C$_s$
170	2.694	4.394	1.347	0.371	7.169	3.341	C$_s$
180	2.966	4.530	1.483	0.337	6.920	1.924	C$_{\infty V}$

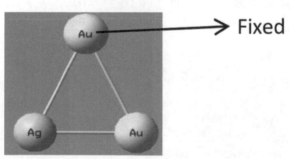

FIGURE 1.2 Triangular structure of cluster AgAu$_2$ with Au atom fixed at the center.

Table 1.2 reveals that least HOMO-LUMO energy gap is obtained at dihedral angle 80° and maximum HOMO-LUMO gap is obtained at dihedral angles varying from 90°–170°. It indicates that an instant cluster is less stable at dihedral angle 80°, and maximum stability is achieved between 90°–170°.

c. In Figure 1.3, one Ag atom is fixed at the center, and two Au atoms placed at the corners are not fixed.

TABLE 1.2   Computed DFT-Based Descriptors of Triangular Structure of Cluster AgAu$_2$ with Au Atom Fixed at the Center

Angles (in Degrees)	HOMO-LUMO Gap (eV)	Electro-negativity (eV)	Hardness (eV)	Softness (eV)	Electro-philicity Index (eV)	Dipole Moment (Debye)	Symmetry
20	1.605	4.258	0.803	0.623	11.295	3.223	C$_s$
30	1.605	4.258	0.803	0.623	11.295	3.222	C$_s$
40	1.605	4.258	0.803	0.623	11.295	3.224	C$_s$
50	1.605	4.258	0.803	0.623	11.295	3.226	C$_s$
60	1.687	4.218	0.844	0.593	10.544	3.752	C$_s$
70	1.605	4.258	0.803	0.623	11.295	3.225	C$_s$
80	1.415	2.503	0.707	0.707	4.429	0.782	C$_s$
90	2.694	4.394	1.347	0.371	7.169	3.341	C$_s$
100	2.694	4.394	1.347	0.371	7.169	3.341	C$_s$
110	2.694	4.394	1.347	0.371	7.169	3.342	C$_s$
120	2.694	4.394	1.347	0.371	7.169	3.341	C$_s$
130	2.694	4.394	1.347	0.371	7.169	3.341	C$_s$
140	2.694	4.394	1.347	0.371	7.169	3.342	C$_s$
150	2.694	4.394	1.347	0.371	7.169	3.344	C$_s$
160	2.694	4.394	1.347	0.371	7.169	3.342	C$_s$
170	2.694	4.394	1.347	0.371	7.169	3.342	C$_s$

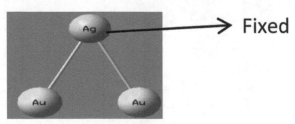

Figure 1.3. Cluster AgAu$_2$ with Ag atom fixed at the center.

Table 1.3 and Figure 1.3 demonstrate that maximum HOMO-LUMO energy gap is obtained at the dihedral angle 30°, and it is constant till 170°; however, there is an exception at 50°. The instant cluster has the least energy gap at 20°.

Table 1.3 Computed DFT-Based Descriptors of Cluster AgAu$_2$ with Ag Atom Fixed at the Center

Angles (in Degrees)	HOMO-LUMO Gap (eV)	Electro-negativity (eV)	Hardness (eV)	Softness (eV)	Electrophilicity Index (eV)	Dipole Moment (Debye)	Symmetry
20	0.163	4.218	0.082	6.125	108.953	0.125	C$_{2v}$
30	1.687	4.218	0.844	0.593	10.544	3.75	C$_{2v}$
40	1.687	4.218	0.844	0.593	10.544	3.751	C$_{2v}$
50	1.605	4.258	0.803	0.623	11.295	3.224	C$_{2v}$
60	1.687	4.218	0.844	0.593	10.544	3.752	C$_{2v}$
70	1.687	4.218	0.844	0.593	10.544	3.757	C$_{2v}$
80	1.687	4.218	0.844	0.593	10.544	3.75	C$_{2v}$
90	1.687	4.218	0.844	0.593	10.544	3.751	C$_{2v}$
100	1.687	4.218	0.844	0.593	10.544	3.751	C$_{2v}$
110	1.687	4.218	0.844	0.593	10.544	3.75	C$_{2v}$
120	1.687	4.218	0.844	0.593	10.544	3.752	C$_{2v}$
130	1.687	4.218	0.844	0.593	10.544	3.751	C$_{2v}$
140	1.687	4.218	0.844	0.593	10.544	3.751	C$_{2v}$
150	1.687	4.218	0.844	0.593	10.544	3.751	C$_{2v}$
160	1.687	4.218	0.844	0.593	10.544	3.751	C$_{2v}$
170	1.687	4.245	0.844	0.593	10.680	3.75	C$_{2v}$
180	0.980	2.394	0.490	1.021	5.853	2.59	D$_{\infty h}$

d. In Figure 1.4, the triangular structure of AgAu$_2$ is formed with an Ag atom fixed at the center and Au atoms located at the corners of the cluster are not fixed. The computed physicochemical properties of all the conformations are presented in Table 1.4 with the corresponding dihedral angles.

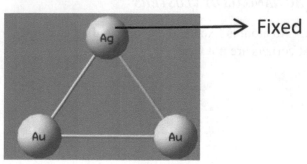

FIGURE 1.4 Triangular structure of cluster AgAu$_2$ with Ag atom fixed at the center.

TABLE 1.4   Computed DFT-Based Descriptors of Triangular Structure of Cluster AgAu$_2$ with Ag Atom Fixed at the Center

Angles (in Degrees)	HOMO-LUMO Gap (eV)	Electro-negativity (eV)	Hardness (eV)	Softness (eV)	Electro-philicity Index (eV)	Dipole Moment (Debye)	Symmetry
20	0.163	4.218	0.082	6.125	108.953	0.125	C$_{2v}$
30	1.687	4.218	0.844	0.593	10.544	3.751	C$_{2v}$
40	1.687	4.218	0.844	0.593	10.544	3.75	C$_{2v}$
50	1.605	4.258	0.803	0.623	11.295	3.224	C$_{2v}$
60	1.687	4.218	0.844	0.593	10.544	3.752	C$_{2v}$
70	1.687	4.218	0.844	0.593	10.544	3.757	C$_{2v}$
80	1.687	4.218	0.844	0.593	10.544	3.747	C$_{2v}$
90	1.687	4.218	0.844	0.593	10.544	3.748	C$_{2v}$
100	1.687	4.218	0.844	0.593	10.544	3.751	C$_{2v}$
110	1.687	4.218	0.844	0.593	10.544	3.751	C$_{2v}$
120	1.687	4.218	0.844	0.593	10.544	3.751	C$_{2v}$
130	1.687	4.218	0.844	0.593	10.544	3.751	C$_{2v}$
140	1.687	4.218	0.844	0.593	10.544	3.749	C$_{2v}$
150	1.687	4.218	0.844	0.593	10.544	3.75	C$_{2v}$
160	1.687	4.218	0.844	0.593	10.544	3.749	C$_{2v}$
170	1.687	4.218	0.844	0.593	10.544	3.751	C$_{2v}$

Table 1.4 and Figure 1.4 demonstrate that maximum HOMO-LUMO energy gap is obtained at the dihedral angle 30°, and it is constant till 170°; however, an exception is marked at 50°. The instant cluster has a minimum HOMO-LUMO energy gap at 20°.

1.3.2   AG$_2$AU NANOALLOY CLUSTERS

In Figure 1.5, Au atom is fixed at the center and other species, i.e., Ag and Au placed at corners are not fixed.

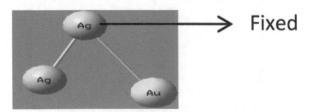

FIGURE 1.5   Cluster Ag$_2$Au with Ag atom fixed at the center.

The orbital energies in the form of HOMO-LUMO gap along with computed DFT-based descriptors viz. electronegativity, hardness, softness, electrophilicity index and dipole moment have been analyzed as a function of dihedral angles and reported in Table 1.5. Figure 1.5 and Table 1.5 establish that the preferred conformation of the instant cluster has maximum HOMO-LUMO energy gap at the dihedral angles varying between 160°–180°, while the least gap is obtained at dihedral angles varying between 20°–110°.

TABLE 1.5 Computed DFT-Based Descriptors of Cluster Ag_2Au with Ag Atom Fixed at the Center

Angles (in Degrees)	HOMO-LUMO Gap (eV)	Electro-negativity (eV)	Hardness (eV)	Softness (eV)	Electro-philicity Index (eV)	Dipole Moment (Debye)	Symmetry
30	1.388	3.850	0.694	0.721	10.682	2.439	C_s
40	1.388	3.850	0.694	0.721	10.682	2.351	C_s
50	1.388	3.850	0.694	0.721	10.682	2.44	C_s
60	1.388	3.850	0.694	0.721	10.682	2.44	C_s
70	1.388	3.850	0.694	0.721	10.682	2.442	C_s
80	1.388	3.850	0.694	0.721	10.682	2.44	C_s
90	1.388	3.850	0.694	0.721	10.682	2.44	C_s
100	1.388	3.850	0.694	0.721	10.682	2.44	C_s
110	1.388	3.850	0.694	0.721	10.682	2.44	C_s
120	2.857	4.095	1.429	0.350	5.870	3.116	C_s
130	2.966	4.122	1.483	0.337	5.730	3.244	C_s
140	2.966	4.122	1.483	0.337	5.730	3.253	C_s
150	2.911	4.095	1.456	0.343	5.760	3.158	C_s
160	3.374	4.218	1.687	0.296	5.272	3.844	C_s
170	3.374	4.190	1.687	0.296	5.204	3.792	C_s
180	3.374	4.286	1.673	0.299	5.488	2.272	$C_{\infty v}$

In Figure 1.6, the triangular structure of Ag_2Au cluster is formed with fixed Au atom at the center, and other Au and Ag atoms are placed at the corners of the triangle. The computed physicochemical properties of all the generated conformations are shown in Table 1.6 with the corresponding dihedral angles.

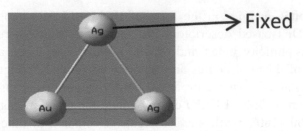

FIGURE 1.6   Triangular structure of cluster Ag_2Au with Ag atom fixed at the center.

Table 1.6 reveals that least HOMO-LUMO energy gap is obtained at dihedral angle 110° and maximum HOMO-LUMO gap is obtained at dihedral angles varying from 130°–170°. It indicates that an instant cluster is less stable at dihedral angle 110°, and maximum stability is achieved between 130°–170°.

TABLE 1.6   Computed DFT-Based Descriptors of Triangular Structure of Cluster Ag_2Au with Ag Atom Fixed at the Center

Angles (in Degrees)	HOMO-LUMO Gap (eV)	Electro-negativity (eV)	Hardness (eV)	Softness (eV)	Electro-philicity Index (eV)	Dipole Moment (Debye)	Symmetry
20	1.388	3.850	0.694	0.721	10.682	2.441	C_s
30	1.442	3.850	0.721	0.693	10.279	2.441	C_s
40	1.415	3.864	0.707	0.707	10.551	2.442	C_s
50	1.388	3.850	0.694	0.721	10.682	2.441	C_s
60	1.388	3.850	0.694	0.721	10.682	2.44	C_s
70	1.388	3.850	0.694	0.721	10.682	2.44	C_s
80	1.388	3.850	0.694	0.721	10.682	2.44	C_s
90	1.415	3.864	0.707	0.707	10.551	2.44	C_s
100	1.388	3.850	0.694	0.721	10.682	2.44	C_s
110	1.361	3.837	0.680	0.735	10.819	2.341	C_s
120	2.803	4.095	1.401	0.357	5.984	3.077	C_s
130	2.966	4.122	1.483	0.337	5.730	3.254	C_s
140	2.966	4.122	1.483	0.337	5.730	3.249	C_s
150	2.966	4.122	1.483	0.337	5.730	3.277	C_s
160	2.966	4.122	1.483	0.337	5.730	3.271	C_s
170	2.966	4.122	1.483	0.337	5.730	3.248	C_s

In Figure 1.7, one Au atom is fixed at the center, and two Ag atoms placed at the corners are not fixed.

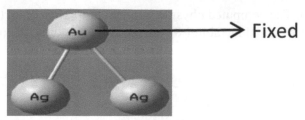

FIGURE 1.7 Cluster Ag_2Au with Au atom fixed at the center.

Table 1.7 and Figure 1.7 demonstrate that maximum HOMO-LUMO energy gap is obtained at the dihedral angle 20°, and the minimum energy gap is obtained at dihedral angles 40° and 70°. The instant cluster has constant HOMO-LUMO energy gap, i.e., 2.367 eV between dihedral angles 90°–170°; however, an exception is marked at angle 150°.

TABLE 1.7 Computed DFT-Based Descriptors of Cluster Ag_2Au with Au Atom Fixed at the Center

Angles (in Degrees)	HOMO-LUMO Gap (eV)	Electro-negativity (eV)	Hardness (eV)	Softness (eV)	Electro-philicity Index (eV)	Dipole Moment (Debye)	Symmetry
20	3.211	3.320	1.605	0.311	3.432	0.674	C_s
30	2.204	3.850	1.102	0.454	6.726	1.525	C_s
40	1.388	3.850	0.694	0.721	10.682	2.392	C_s
50	1.442	3.850	0.721	0.693	10.279	2.428	C_s
60	1.442	3.850	0.721	0.693	10.279	2.417	C_s
70	1.388	3.850	0.694	0.721	10.682	2.441	C_s
80	1.415	3.837	0.707	0.707	10.403	2.395	C_s
90	2.367	3.877	1.184	0.422	6.351	1.146	C_s
100	2.367	3.877	1.184	0.422	6.351	1.147	C_s
110	2.367	3.877	1.184	0.422	6.351	1.146	C_s
120	2.367	3.877	1.184	0.422	6.351	1.145	C_s
130	2.367	3.877	1.184	0.422	6.351	1.146	C_s
140	2.367	3.877	1.184	0.422	6.351	1.145	C_s
150	2.422	3.877	1.211	0.413	6.208	1.098	C_s
160	2.367	3.877	1.184	0.422	6.351	1.144	C_s
170	2.367	3.877	1.184	0.422	6.351	1.145	C_s
180	2.531	3.905	1.265	0.395	6.025	1.598	$D_{\infty h}$

In Figure 1.8, the triangular structure of Ag_2Au is formed with an Ag atom fixed at the center and Au atoms located at the corners of the cluster are not fixed. The computed physicochemical properties of all the conformations are presented in Table 1.8 with the corresponding dihedral angles.

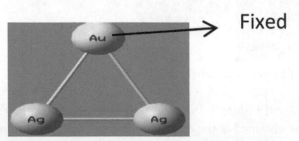

FIGURE 1.8   Triangular structure of cluster Ag_2Au with Au atom fixed at the center.

TABLE 1.8   Computed DFT-Based Descriptors of Triangular Structure of Cluster Ag_2Au with Au Atom Fixed at the Center

Angles (in Degrees)	HOMO-LUMO Gap (eV)	Electro-negativity (eV)	Hardness (eV)	Softness (eV)	Electro-philicity Index (eV)	Dipole Moment (Debye)	Symmetry
20	1.388	3.905	0.694	0.721	10.987	1.637	C_{2v}
30	1.388	3.905	0.694	0.721	10.987	1.637	C_{2v}
40	1.388	3.905	0.694	0.721	10.987	1.638	C_{2v}
50	1.388	3.905	0.694	0.721	10.987	1.637	C_{2v}
60	1.388	3.905	0.694	0.721	10.987	1.636	C_{2v}
70	1.388	3.905	0.694	0.721	10.987	1.636	C_{2v}
80	2.359	3.909	1.180	0.424	6.476	0.627	C_{2v}
90	2.359	3.909	1.180	0.424	6.476	0.626	C_{2v}
100	2.359	3.909	1.180	0.424	6.476	0.627	C_{2v}
110	2.359	3.909	1.180	0.424	6.476	0.628	C_{2v}
120	2.359	3.909	1.180	0.424	6.476	0.627	C_{2v}
130	2.359	3.909	1.180	0.424	6.476	0.627	C_{2v}
140	2.359	3.909	1.180	0.424	6.476	0.626	C_{2v}
150	2.359	3.909	1.180	0.424	6.476	0.627	C_{2v}
160	2.359	3.909	1.180	0.424	6.476	0.628	C_{2v}
170	2.359	3.909	1.180	0.424	6.476	0.628	C_{2v}

Table 1.8 and Figure 1.8 demonstrate that maximum HOMO-LUMO energy gap is obtained at dihedral angle 80°, and it is constant till 170°. The instant cluster has a minimum HOMO-LUMO energy gap at dihedral angles varying between 20°–70°.

1.4 CONCLUSION

In this chapter, we have studied the conformational effect of AgAu nanoalloy clusters on the basis of DFT platform. We have computed CDFT-based descriptors viz., molecular hardness, softness, electronegativity, electrophilicity index, dipole moment and symmetry along with HOMO-LUMO energy gap of trimer AgAu clusters. The result reveals that the linear structure of AgAu nanoalloy clusters is more stable than other conformations. The reason for the stability of small linear structures may be understood from the molecular orbital concepts. However, an exception has been observed in the case of Au_2Ag (where Ag atom is fixed). In this report, we have observed large HOMO-LUMO gap for linear structure, except $AgAu_2$ (where Ag atom is fixed at the center). In the case of $AgAu_2$ cluster, where the Ag atom is fixed at the center, maximum HOMO-LUMO gap is observed at dihedral angle 30°. The computed HOMO-LUMO energy gap of these clusters runs hand in hand, along with their computed hardness values.

KEYWORDS

- **Ag-Au**
- **conformational analysis**
- **density functional theory**
- **dipole moment**
- **HOMO-LUMO gap**

REFERENCES

1. Ghosh, D. C., (2006). A Quest for the origin of barrier to the internal rotation of hydrogen peroxide (H_2O_2) and fluorine peroxide (F_2O_2), *Int. J. Mol. Sci., 7*, 289.
2. Freitas, M. P. D., & Ramalho, T. D. C., (2013). Ramalho, Employing conformational analysis in the molecular modeling of agrochemicals: insights on QSAR parameters of 2, 4-D, *Ciênc. Agrotec., 37*, 485.
3. Castro, A. T., & Figuero-Villar, J. D., (2002). Molecular structure, conformational analysis and charge distribution of pralidoxime: Ab initio and DFT studies, *Int. J. Quantum Chem., 89*, 135.
4. Gerbst, A. G., Nikolaev, A. V., Yashunsky, D. V., Shashkov, A. S., Dmitrenok, A. S., & Nifantiev, N. E., (2017). Theoretical and NMR-based conformational analysis of phosphodiester-linked disaccharides, *Scientific Reports, 7*.

5. Alvarez-Ros, M. C., & Palafox, M. A., (2014). Conformational analysis, molecular structure and solid state simulation of the antiviral drug acyclovir (zovirax) using density functional theory methods, *Pharmaceuticals, 7*, 695.

6. Krishnamurty, S., Stefanov, M., Mineva, T., Begu, S., Devoisselle, J. M., Goursot, A., Zhu, R., & Salahub, D. R., (2008). Density functional theory-based conformational analysis of phospholipid molecule (dimyristoyl phosphatidylcholine), *J. Phys. Chem. B, 112*, 13433.

7. Bao, G., Kamm, R. D., Thomas, W., Hwang, W., Fletcher, D. A., Grodzinsky, A. J., Zhu, C., & Morfad, M. R. K., (2010). Molecular biomechanics: the molecular basis of how forces regulate cellular function, *Mol. Cell Biomech., 3*, 91.

8. Agrawal, M., Kumar, A., & Gupta, A., (2017). Conformational stability, spectroscopic signatures and biological interactions of proton pump inhibitor drug lansoprazole based on structural motifs, *RSC Adv., 7*, 41573.

9. Mary, Y. S., Raju, K., Panicker, C. Y., Al-Saadi, A. A., Thiemann, T., & Alsenoy, C. V., (2014). Molecular conformational analysis, vibrational spectra, NBO analysis and first hyperpolarizability of (2E)-3-phenylprop-2-enoic anhydride based on density functional theory calculations, *Spectrochim. Acta, Part A, 128*, 638.

10. Srivastava, A. K., Dwivedi, A., Kumar, A., Gangwar, S. K., Misra, N., & Sauer, S. P. A., (2016). Conformational analysis, inter-molecular interactions, electronic properties and vibrational spectroscopic studies on cis-4-hydroxy-d-proline, *Cogent Chem., 2*, 1149927.

11. Sarmah, P., & Deka, R. C., (2009). DFT-based QSAR and QSPR models of several cis-platinum complexes: solvent effect, *J. Comput. Aided Mol. Des., 23*, 343.

12. Pitzer, R. M., (1983). The barrier to internal rotation in ethane, *Acc. Chem. Res., 16*, 207.

13. Schleyer, P. V. R., Kaupp, M., Hampel, F., Bremer, M., & Mislow, K., (1992). Relationships in the rotational barriers of all Group 14 ethane congeners H3X-YH3 (X,Y = C, Si, Ge, Sn, Pb) Comparisons of ab initio pseudopotential and all-electron results, *J. Am. Chem. Soc., 114*, 6791.

14. Ghosh, D. C., & Rahman, M. A., (1996). Study of the origin of barrier to internal rotation of molecules, *Chem. Environ. Res., 5*, 73.

15. Freeman, F., Tsegai, Z. M., Kasner, M. L., & Hehre, W., (2000). A comparison of the ab initio calculated and experimental conformational energies of alkylcyclohexanes, *J. Chem. Educ., 77*, 661.

16. Hafner, J., Wolverton, C., & Ceder, G., (2006). Toward computational materials design: the impact of density functional theory on materials research, *MRS Bulletin, 31*, 659.

17. Wang, H. Q., Kuang, X. Y., & Li, H. F., (2010). Density functional study of structural and electronic properties of bimetallic copper-gold clusters: comparison with pure and doped gold clusters, *Phys. Chem. Chem. Phys., 12*, 5156.

18. Becke, A. D., (1993). Density functional thermochemistry. III. The role of exact exchange, *J. Chem. Phys., 98*, 5648.

19. Lee, C., Yang, W., & Parr, R. G., (1988). Development of the colle-salvetti correlation-energy formula into a functional of the electron density, *Phys. Rev. B: Condens. Matter, 37*, 785.

20. Mielich, B., Savin, A., Stoll, H., & Preuss, H., (1989). Results obtained with the correlation energy density functional of Becke and Lee, yang and Parr, *Chem. Phys. Lett., 157*, 200.
21. Jiang, Z. Y., Lee, K. H., Li, S. T., & Chu, S. Y., (2006). Structures and charge distributions of cationic and neutral Cun-1Ag clusters ($n = 2$–8), *Phys. Rev. B, 73*, 235423.
22. Zhao, Y. R., Kuang, X. Y., Zheng, B. B., Li, Y. F., & Wang, S. J., (2011). Equilibrium geometries, stabilities and electronic properties of the bimetallic M2-doped Aun (M = Ag, Cu; $n = 1$–10) clusters: comparison with pure gold clusters, *J. Phys. Chem. A, 115*, 569.
23. Gaussian 03, Revision, C.02, Frisch, M. J., Trucks, G. W., Schlegel, H. B., Scuseria, G. E., Robb, M. A., Cheeseman, J. R., et al., (2004). Gaussian, Inc., Wallingford CT.
24. Koopmans, T., (1934). Ordering of wave functions and eigenenergies to the individual electrons of an atom, *Physica, 1*, 104.
25. Parr, R. G., & Yang, W., (1989). *Density Functional Theory of Atoms and Molecules*. Oxford University Press, Oxford.

CHAPTER 2

Methods for Determination of Nanostructures Mechanical Properties

A. A. SHUSHKOV and A. V. VAKHRUSHEV

Udmurt Federal Research Center of the Ural Branch of the Russian Academy of Sciences, Institute of Mechanics, Department of Mechanics of Nanostructures, Izhevsk, Russia, E-mail: vakhrushev-a@yandex.ru

ABSTRACT

The chapter provides an overview of methods for determining the mechanical properties of nanostructures: calculation and experimental methods, methods based on mathematical computer simulation and methods based on their joint application. The features and deficiencies of each method are shown. In many papers of nanostructures mechanical properties study increase the elastic modulus with a decrease in size is tended. However, there are also opposite results. Since the results have an ambiguous and sometimes contradictory character of the dependence of the elastic modulus from the of nanostructures size, investigations in the field of developing a generalized, precise method for determining the mechanical properties of nanostructures of arbitrary geometric shape are relevant.

2.1 INTRODUCTION

In recent decades, many methods for determining the mechanical properties (statistical modulus of Young's elasticity, Poisson's ratio, and others) of arbitrary geometric shape nanostructures (nanofibers, nanotubes, nanoparticles) have been developed [1–5]. The emergence of the creation direction of similar techniques is associated with the phenomenon of a change (in most cases and increase) in the mechanical characteristics of nanostructures with a decrease in their size [6–8]. Particularly great interest in their research has appeared in connection with the creation of nanocomposite materials with improved predicted mechanical properties, which is one of the priority directions of research on a global scale [9].

So far, there are no precise methods for determining the elastic modulus and the Poisson's ratio of nanostructures. All existing methods have a number of drawbacks. In particular, a direct experimental measurement of the nanostructures elastic characteristics is technically complex cost-effective and, in some cases, not at all feasible because of the "small" size (several tens of nanometers and sometimes angstroms) of nanostructures various forms. In these conditions, numerical computer simulation is the most acceptable way of studying the nanoparticles elastic properties. Computational modeling is an alternative and promising way of establishing the characteristics of nanoparticles.

In addition, the methods use of computer mathematical modeling makes it possible to predict the real mechanical properties of nanostructures as well as composites based on them with the least time and without the use of expensive equipment.

The practical utility of the study is that it is related to the dependence calculation of the elastic modulus versus the nanoparticles size, which will allow providing the production of nanocomposite materials with established, required elastic properties. The presence of such dependence significantly reduces the volume of produced experimental research by obtaining information about the mechanical characteristics of nanoparticles.

Solving the problems of eliminating the shortcomings of the determination of experimental methods, which are presented below, led to the need to create analytical models to find the mechanical properties of nanostructures. A necessary part of each model of this kind is the correspondence that the elastic characteristics of nanostructures coincide with their values obtained from macro experiments [10]. The results of these methods of calculating the nanoparticles elastic modulus on their size obtained by different authors are ambiguous and sometimes even contradictory character [6, 11]. This is due to the choice of a different model of investigated nanoparticles, the computer simulation method (finite element method (FEM), molecular dynamics method, quantum mechanics method). And also, recent investigations show that in the field of nanoparticle size (diameter) less than ≈ 6 nm under using the same model of the nanostructure, the same method of mathematical modeling the mechanical properties of nanoparticles depend on the method of nanoobject deformation [12]. This leads to the conclusion that the use of analytical models obtained on the basis of the results from macro experiments is incorrect. In this connection, the task arises – the development of a generalized, precise method for determining the nanoparticles mechanical properties of an arbitrary shape based on a comparison of the results obtained by experimental methods and various methods of mathematical computer simulation in order to obtain a more accurate result.

All methods of determining the nanostructures mechanical properties can be conditionally divided into two types (Figure 2.1).

"Direct" definition of the nanostructure mechanical properties based on its deformation. The deformation of the composite material and the subsequent "indirect" calculation of the nanostructures mechanical properties of the same size [13] included in its composition. The second method seems simpler since the sample is massive. However, when using computer simulation methods require a longer time for calculation. Therefore, the "direct" definition of the nanostructures mechanical properties is a topical task that allows predicting the properties of composite material on their basis, to reduce the costs and time of creating a composite material with the required, necessary for the particular case improved mechanical properties.

Another urgent task of this kind is to find the transition, the size of the nanoparticle (critical diameter) after which the properties of the nanoparticles will coincide with the properties of the macro material. In this area, there are only a few research papers [14].

We give a brief review of the methods for determining the mechanical characteristics of micro- and nanoparticles of materials.

2.2 EXPERIMENTAL CALCULATION METHODS FOR DETERMINING THE NANOSTRUCTURES MECHANICAL PROPERTIES

One of the main methods for investigation of the mechanical properties of thin films is the nanoindentation method. A new task of research by

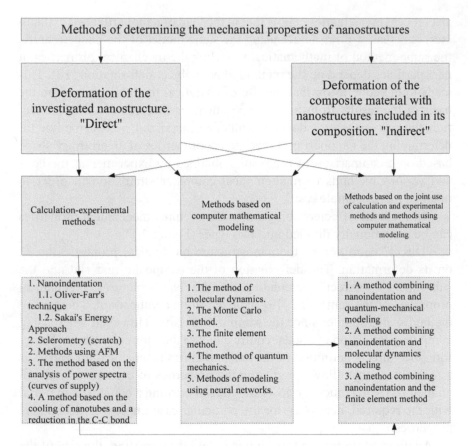

FIGURE 2.1   Methods of determining the nanostructures mechanical characteristics.

the nanoindentation method is the determination of the nanoparticle's mechanical characteristics.

The study of mechanical properties based on the pressing of the indenter is widely used in the investigation of films and surface layers [15–18]. There is much information about the elastic modulus, the hardness obtained by this method [19–33]. The variety of tests types by continuous nanoindentation has become a common use for measuring the mechanical properties of materials.

For these investigations, a wide variety of tested devices with indenters of various forms, working with sizes ranging from nanoscale to macroscale have been developed. A common feature of these tests is that the applied load is read as a function of the indentation depth during the application mode and the removal of the load. A significant advantage of this method consists in the accuracy with which the elastic modulus is measured. However, since the indenter is relatively small in relation to the surface area, all measurements are local. In addition, the elastic modulus is determined in the surface of the nanomaterial sample. What will be the elastic modulus within the sample to be determined on the basis of this method is not possible. The nanoindentation method is applicable for determining the elastic modulus in thin films and is not applicable for comparatively small bulk nanoparticles. There are several difficulties in applying this method connected with the fact that the results of measurements do not always correspond to the true characteristics of the investigated samples.

The reason for this first of all is the influence of the substrate on which the particle is applied and which as a rule has completely different mechanical properties. In addition, there is the problem of pressing or conversely extrusion of the test material along with the faces of the indenter pyramid in the foreign literature – pile-up or sink-in which leads to inaccuracies in determining the contact area of the indenter with the sample and as a result distortion of the results. Also, the measurement result depends on the method of analysis of the indentation curves. Despite a considerable number of works in this field, the problem of measuring the mechanical characteristics of thin films, nano-, and microparticles by the method of nanoindentation has not yet been solved.

The nanoindentation method is used in particular to determine the elastic modulus in epitaxial layers of gallium nitride [34]. The technique is based on the solution of the Hertz task for the elastic pressing of a steel sphere into the investigated surface – an elastic isotropic body. Therefore, for anisotropic materials, only approximate values can be obtained by this method. It is established that the isotropic approximation used here is justified. However, the isotropy in tasks of this kind requires a more detailed justification, which is very important.

Most of the results for calculating the mechanical characteristics by the nanoindentation method are determined based on the Oliver-Farr technique, which is described in detail [35]. However, this technique has a high error under determining the mechanical properties of "soft" materials require a lot of very accurate calibrations of the used equipment (calibration of the penetration depth of the indenter, load, system compliance, projection of the diamond tip area, etc.). Failure to perform at least one of the calibrations will lead to an inaccurate determination of the mechanical characteristics of the investigated surface.

In addition, under determining the mechanical characteristics, it is necessary to take into account the influence of the piles depressions. Non-considering of these phenomena can lead to a "large" error in determining the reduced elastic modulus, and hardness up to 50% [36]. Considering all the above, there is a need to create new methods for determining the mechanical characteristics of thin films micro- and nanostructures.

The method for determining the reduced elastic modulus is presented in the thesis [36]. The original method for measuring the elastic modulus was developed by analyzing the power spectra (supply curves) [37, 38]. In the process of measuring the indenter oscillates in the direction normal to the surface of the sample with an amplitude of ~ 5 nm and a frequency of ~ 10 kHz. During the measurement, the indenter moves in steps (~ 0.1 nm) by means of a piezoceramic actor (scanner) towards the surface. At the moment of contact with the surface as a result of the interaction of the tip with the material, the amplitude and frequency of the indenter oscillation are changed. In the presented technique, the measured parameter is the frequency. The change of frequency depends on the characteristics of the indenter as well as from the elastic properties of the material in the contact region. During the experiment, the frequency change is recorded for each position of the piezo scanner. The obtained dependence of the change of the indenter frequency from the displacement of the piezo scanner is called the force spectrum or the approach curve.

To interpret the obtained results and substantiate the possibility of quantitative measurements of the elastic modulus using the described technique a model describing the change in the oscillation frequency of an indenter as a function of the displacement of indenter is proposed. The model is based on the analytical solution of the "Hertz task" on the mutual deformation of two solid spheres during their compression, which is described in detail in [39]. The use of the Hertz model to describe the interaction of the indenter and the sample, in this case, is justified since the surface forces are negligibly small in comparison with the total interaction forces.

The method developed by the authors for determining the mechanical characteristics is preferable to the method of nanoindentation since the measurement takes place on the elastic portion of the indenter supply the depth of immersion of the tip is small the measurement can be carried out with any roughness of the surface which is very important. The thickness of the thin film can be 100 nm. The elastic contact corresponds to the immersion of the indenter at 5–15 nm.

It should be specially emphasized that the presented methods of measurement and subsequent calculation of mechanical properties are applicable only for thin films, individual components in complex multiphase structures, and near-surface layers of materials. As applied to nanoparticles, they are not always acceptable.

However, experimental methods for determining the mechanical properties of nanostructures exist, and almost all of them are the second type. Here are several methods to determine the mechanical characteristics of nanoobjects.

It should be noted that there are theoretical [40] as well as experimental methods for determining the nanotubes elastic modulus for example by vibrations amplitude value of isolated single-layered and multilayered nanotubes [41, 42] which were studied by a transmission electron microscope. In another experimental method, a multilayered nanotube was attached to the substrate by ordinary lithography the force was applied and measured at different distances from the attachment point by an atomic force microscope. Ref. [43] determine the elastic modulus, a suspension of single-walled nanotubes was passed through the membrane nanotubes were suspended in pores, and their deflections were measured by an atomic force microscope.

There is a method for determining the elastic modulus of carbon nanotubes or carbon fibers added to an epoxy matrix [44] based on the fact that a sample of the material is exposed to a HeN laser. Obtain the Raman spectrum of the sample material using Renishaw Ramascope. Cool the sample to a certain temperature using the Linkam THMS 600 cooling cell, thereby causing deformation of the composite material. Cooling is performed by injecting nitrogen to obtain the Raman spectrum of the sample material. The displacement of the intensity peak of the sample after cooling is due to an axial decrease in the length of the C-C bond. Calculate the elastic modulus of carbon nanotubes or carbon fibers.

Nanocomposites [6] consisting of a polymeric matrix – polymethylmethacrylate (PMMA) with the addition of a specified percentage of quartz nanoparticles have been experimentally studied. To determine the elastic modulus of nanoparticles, an inverse analysis of the method of equivalent inclusions was proposed. Using numerical analysis, the elastic modulus for quartz particles was calculated.

It was found that the elastic modulus of the nanocomposite remained almost constant with a volume fraction of the quartz particle fraction of 8% and increased significantly when the particle sizes were nano-ordered. This result was confirmed and compared with a three-phase model. It is shown that the proposed method is effective for predicting the elastic modulus of inorganic particles in nanocomposites.

2.3 METHODS OF DETERMINING MECHANICAL PROPERTIES OF NANOSTRUCTURES USING COMPUTER MODELING TECHNIQUES

Many methods for determining the elastic modulus and the Poisson's ratio of nanostructures are based on the use of computer mathematical modeling techniques. The mechanical characteristics of particles are determined on the basis of modeling the deformation of the composite material and the subsequent indirect calculation of the elastic modulus of particles included in its composition [45, 46]. In particular, in Ref. [47] using computer simulations, the Monte Carlo method was used to study samples consisting of titanium nanoparticles aggregates. The "equivalent" titanium model is deformed along one of the axes. The elastic modulus is calculated from the deformation energy of the modeled sample by the FEM. The elastic modulus of a sample consisting of titanium nanoparticles increases with decreasing particle size. It was found that the elastic modulus of the sample studied depends on the size of the nanoparticles and does not depend on the size of the aggregates of the nanoparticles.

However, it was shown in [11] that elastic modulus of composite materials increases with increasing radius of quartz nanoparticles. The volume fraction of the content of quartz nanoparticles was 5%. The calculations were carried out using the molecular dynamics method using the "equivalent" three-phase continuous model. The results of the calculations are compared with the two Mori-Tanaka phase models. It can be seen that the elastic modulus of composite materials remains constant with the use of the Mori-Tanaka model.

The theoretical basis of the increasing tendency of the elastic modulus with a decrease of the nanocrystals dimensions is presented in the work of Russian scientists [10] using the example of a two-dimensional strip made of a single-crystal material with a hexagonal close-packed lattice. The results agree well with the experimental data obtained [48]. It was found that the elastic modulus can be two times different from its macroscopic value. It is shown that the size and shape of the nanocrystal introduce additional anisotropy into its mechanical properties.

However, verification is necessary for other types of crystal lattices, especially three-dimensional ones. In the next paper [49], a three-dimensional nanocrystal with a cubic face-centered lattice was investigated. It is found that the elastic modulus increases 3.83 times with respect to its macroscopic value.

The values of the elastic modulus do not always increase with decreasing dimensions of the nanocrystals. Also, the authors made a very good conclusion that the concepts of classical continuum mechanics, including the theory of elasticity when applied to nanoobjects, should be used with great caution. It is necessary to consider the change in mechanical characteristics as the scale of the object under consideration approaches the nanometer scale. Particular attention must be paid to quantities that are fundamentally ambiguous at the nanoscale such as the elastic modulus. When using them, it is necessary to clearly define what exactly is meant by the indicated values when applied to nanoobjects. However, all this does not mean that the classical theory of elasticity is not applicable at the nanoscale. Simply it should be used in view of the scale effects, and the adequacy of the continuum approach should be evaluated when considering specific tasks.

A study of the elastic properties of three-dimensional crystals is presented in Ref. [50]. The reaction of small aluminum nanoparticles (from 2 to 13 atoms), silicon (from 5 to 18), and zirconium dioxide (from 18 to 30) to tensile and compression were investigated using quantum-mechanical computer simulation (the theory of the electron density functional and the pseudopotential method). It was found that the elastic modulus of nanoparticles, as a rule, is several times higher than the corresponding values for bulk material. With an increase in the number of atoms in the nanoparticle, the elastic modulus behaves differently for particles with different types of interatomic bonding. In the case of silicon, its value rapidly tends to a value characteristic of a massive material in the case of zirconia practically does not change in the case of aluminum the dependence of the elastic modulus from the nanoparticle size is nonmonotonic and is determined by the geometry of the particle.

In general, it can be said that the elastic characteristics of nanoparticles significantly exceed the corresponding values typical for massive materials however in the case of silicon a rapid convergence of nanoparticle elastic modulus and massive material is observed with an increasing number of atoms in the particle.

The calculated elastic characteristics are in good agreement with the experimental data for zirconia nanoparticles despite the fact that particles of much larger size (18 nm) were investigated in Ref. [51].

There is a technique for calculating the elastic modulus of spherical nanoparticles on the basis of their stretching or compression by concentrated forces applied at opposite ends of the diameter [52]. The method of molecular dynamics is applied. The material of this element of the body is considered homogeneous and isotropic. It should be noted that in the general case, the properties of the nanoelement can be inhomogeneous (possible anisotropy is not considered). This is due to a change in the atomic structure of the nanoelement in its surface area. The elastic modulus is determined by the coincidence of the displacement vectors of atoms from the center to the point of application of the force of the nanoparticle and the equivalent elastic element (ball). Figure 2.2 illustrates the general trend of increasing the elastic modulus; however, for different materials,

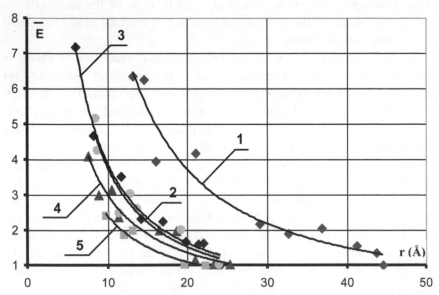

FIGURE 2.2 Dependence of the relative elastic modulus $\bar{E}$ from the radius r (Å) for nanoparticles from various materials: 1 – cesium, 2 – calcium, 3 – zinc, 4-magnesium, and 5 – potassium.

it is not the same. As the relative modulus, the ratio of the elastic modulus to its asymptotic value at the maximum diameter of the nanoparticle is chosen. The relative zinc modulus increases more than 7 times, and the potassium modulus only 2.5 times.

An analogous technique for determining the elastic modulus based on the compression of a nanoparticle and an elastic equivalent element uniformly distributed over the surface by pressure has been developed [12]. The dependence of the elastic modulus on the radius obtained on the basis of the compression of nanoparticles by a uniformly distributed pressure over the surface (for Poisson's ratio 0.3) does not change with increasing cesium nanoparticle size (Figure 2.3). The elastic modulus of cesium nanoparticles calculated by this method is equal to the reference value with a relative error of 5%.

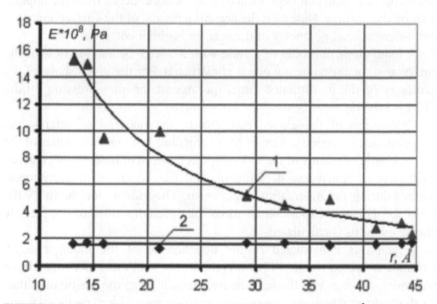

FIGURE 2.3 Dependence of the elastic modulus E, Pa on the radius r, Å of the cesium nanoparticles: 1 – stretching by concentrated forces, and 2 – compression using pressure uniformly distributed over the surface [12].

Thus it can be concluded that the elastic modulus of nanoparticles of "small" sizes (radius less than ≈ 30 Å for cesium) depends on the stress-strain state formed after the application of different types of loading and the isotropic approximation, in this case, is not justified.

2.4   METHODS FOR THE DETERMINING OF MECHANICAL PROPERTIES OF NANOSTRUCTURES BASED ON THE JOINT USE OF EXPERIMENT AND TECHNIQUES OF COMPUTER MODELING

A technique was developed [53] for determining the elastic modulus of nanoparticles of arbitrary shape based on a comparison of the results of numerical calculations by the FEM and experiments. This task of determining the elastic modulus is solved by spraying particles on the substrate experimentally scanning the surface and finding the particle then indenting any indenter into the center of the particle.

The load applied to the indenter is chosen from the condition that the deformation is elastic. Only elastic deformation is considered since at the initial stages of indenter immersion all deformation is reversible. We obtain the experimental dependence of indentation depth from the applied load on the particle. Based on the scanning results of the surface area, the three-dimensional geometry of the surface is obtained.

A finite-element model of a zone with a micro- or nanoparticle of the corresponding experimental one is constructed. For the given shape of the geometry of the investigated zone, the task of indenter pressing by the FEM is solved.

Calculation of the elastic modulus of the material of micro- and nanoparticles is carried out by FEM, proceeding from known values of the shape, elastic constants of the indenter, force, depth of indentation, a given coefficient of transverse deformation and the geometry of the particle surface during the indentation experiment. The indentation depth by the FEM for the same experimental force for different values of the particle elastic modulus is calculated.

A graph is constructed of the dependence of the indenter depth variation obtained by the FEM in comparison with the experimental one ($\Delta h = h_{EXPER} - h_{FEM}$) on the investigated particle from the elastic modulus. At the point at which the dependence reaches zero (the experimental and calculated penetration depths of the indenter coincide), the elastic modulus of the particle under investigation is determined.

2.5   CONCLUSIONS

A review of methods for determining the mechanical properties of near-surface layers, nanostructures (experimental and methods based on mathematical computer simulation) is given. The advantages and disadvantages of each method are presented. An advantageous method for determining mechanical properties in comparison with the method of indentation is the method of power spectroscopy [36].

Using the Monte Carlo method, the method of equivalent inclusions, the method of molecular dynamics, the pseudopotential method, and the theory of the electron density functional, the elastic modulus increases with decreasing nanoparticle size was shown. However, this trend is ambiguous using the method of molecular dynamics using an "equivalent" three-phase continuous model of the study showed that the elastic modulus increases with the size of the nanoparticles and using the Mori-Tanaka model remains constant using the model of the elastic "equivalent" element the elastic modulus increases with decreasing size of the nanoparticles.

Theoretical calculations show that the elastic modulus should increase with decreasing size of nanocrystals. It is shown that using the same method of mathematical computer simulation (the molecular dynamics method), the elastic modulus of "small" nanoparticles (radius ≈ 0.3 nm for cesium nanoparticles) depends on the stress-strain state, so the assumed isotropic approximation is not justified.

ACKNOWLEDGMENTS

The works were carried out with financial support from the Research Program of the Ural Branch of the Russian Academy of Sciences (project 18-10-1-29).

KEYWORDS

- elastic modulus
- finite element method
- mechanical properties
- nanoparticles

REFERENCES

1. Golovin, Y. I., (2009). *Nanoindentation and its Capabilities.* Moscow: Mashinostroenie.
2. Bulychev, S. I., & Alekhin, V. P., (1990). *Testing of Materials by Continuous Indent Pressing.* Moscow: Mashinostroenie.
3. Fedosov, S. A., & Peshek, L., (2004). *Determination of Materials Mechanical Properties by Microindentation: Modern Foreign Methods.* Moscow: Faculty of Physics.
4. Vakhrushev, A. V., Lipanov, A. M., & Shushkov, A. A., (2007). *The Method for Determining the Young's Modulus of Materials Elasticity.* Patent of the Russian Federation No 2292029.
5. Vakhrushev, A. V., Lipanov, A. M., & Shushkov, A. A., (2007). *The Method for Determining the Young's Modulus of Materials Elasticity.* Patent of the Russian Federation No 2296972.
6. Ni, Q. Q., Fu, Y., & Iwamoto, M., (2004). Evaluation of elastic modulus of nano particles in PMMA/silica nanocomposites. *Journal of the Society of Materials Science, 53*(9), 956–961.
7. Ruoff, R. S., & Pugno, N. M., (2004). Strength of nanostructures. *Proceedings of the 21ˢᵗ International Congress of Theoretical and Applied Mechanics,* 303–311.
8. Dingreville, R., Qu, J., & Cherkaoui, M., (2004). Surface free energy and its effect on the elastic behavior of nano-sized particles, wires and films. *Journal of the Mechanics and Physics of Solids, 53*(8), 1827–1854.
9. Ge, W., Ka-Cheung, C., Linli, Z., Ligang, S., & Jian, L., (2017). Dual-phase nanostructuring as a route to high-strength magnesium alloys. *Nature, 545,* 80–83.
10. Krivtsov, A. M., & Morozov, N. F., (2002). On mechanical characteristics of nanocrystals. *Physics of the Solid State, 44*(12), 2260–2265.
11. Odegard, G. M., Clancy, T. C., & Gates, T. S., (2005). Modeling of the mechanical properties of nanoparticle/polymer composites. *Polymer, 46*(2), 553–562.
12. Vakhrushev, A. V., Vakhrusheva, L. L., & Shushkov, A. A., (2011). Numerical analysis of elastic modulus change of crystalline metal nanoparticles under the action of different types of loading. *Izvestiya Tula State University, Natural Sciences, 3,* 137–150.
13. Kristensen, R., (1982). *Introduction to Composite Mechanics.* Moscow: Mir.
14. Vakhrushev, A. A., Fedotov, A. Y., Shushkov, A. A., & Shushkov, A. V., (2011). Modeling of the metal nanoparticles formation, the study of structural, physico-mechanical properties of nanoparticles and nanocomposites. *Izvestiya Tula State University, Natural Sciences, 2,* 241–253.
15. Andrievskiĭ, R. A., Kalinnikov G. V., Hellgren, N., Sandstrom, P., & Shtanskiĭ, D. V., (2000). *Nanoindentation and Strain Characteristics of Nanostructured Boride / Nitride Films. Physics of the Solid State, 42*(9), 1671–1674.
16. Vorobev, V. L., Bykov, P. V., Bayankin, V. Y., Novoselov, A. A., Bureev, O. A., Shushkov, A. A., & Vakhrushev, A. V., (2014). Composition of surface layers of carbon steel depending on accelerating voltage pulse Cr+ ion radiation. *Chemical Physics and Mezoscopy, 16*(2), 257–262.

17. Shojaei, O. R., & Karimi, A., (1998). Comparison of mechanical properties of TiN thin films using nanoindentation and bulge test. *Thin Solid Films, 332*(2), 202–208.

18. Sklenicka, V., Kucharova, K., Pahutova, M., Vidrich, G., Svoboda, M., & Ferkel, H., (2005). Mechanical and creep properties of electrodeposited nickel and its particle-reinforced nanocomposite. *Reviews on Advanced Materials Science, 10*(2), 171–175.

19. Gong, J., Miao, H., & Peng, Z., (2003). A new function for the description of the nanoindentation unloading data. *Scripta Materialia, 49*(1), 93–97.

20. Jung, Y. G., Lawn, B. R., Martyniuk, M., Huang, H., & Hu, X. Z., (2004). Evaluation of elastic modulus and hardness of thin films by nanoindentation. *Journal of Materials Research, 19*(10), 3076–3080.

21. Hua, W., & Wu, X., (2002). Nanohardness and elastic modulus at the interface of TiCx/Ni₃Al composites determined by the nanoindentation technique. *Applied Surface Science, 189*(1/2), 72–77.

22. Lee, Y. H., & Kwon, D., (2003). Measurement of residual-stress effect by nanoindentation on elastically strained (100) W. *Scripta Materialia, 49*(5), 459–465.

23. Qi, H. J., Teo, K. B. K., Lau, K. K. S., Boyce, M. C., Milne, W. I., Robertson, J., & Gleason, K. K., (2003). Determination of mechanical properties of carbon nanotubes and vertically aligned carbon nanotube forests using nanoindentation. *Journal of the Mechanics and Physics of Solids, 51*(11/12), 2213–2237.

24. Maier, P., Richter, A., Faulkner, R. G., & Ries, R., (2002). Application of nanoindentation technique for structural characterization of weld materials. *Materials Characterization, 48*(4), 329–339.

25. Mante, F. K., Baran, G. R., & Lucas, B., (1999). Nanoindentation studies of titanium single crystals. *Biomaterials, 20*(11), 1051–1055.

26. Cho, S. J., Lee, K. R., Eun, K. Y., Hahn, J. L., & Ko, D. H., (1999). Determination of elastic modulus and Poisson's ratio of diamond-like carbon films. *Thin Solid Films, 341*(1/2), 207–210.

27. Sriram, S., & Bharat, B., (2002). Development of AFM-based techniques to measure mechanical properties of nanoscale structures. *Sensors and Actuators A: Physical, 101*(3), 338–351.

28. Gouldstone, A., Kon, H. J., Giannakopoulos, A. E., & Suresh, S., (2000). Discrete and continuous deformation during nanoindentation of thin films. *Acta Materialia, 48*(9), 2277–2295.

29. Giddings, V. L., Kurtz, S. M., Jewett, C. W., Foulds, J. R., & Edidin, A. A., (2001). A small punch test technique for characterizing the elastic modulus and fracture behavior of PMMA bone cement used in total joint replacement. *Biomaterials, 22*(3), 1875–1881.

30. Vaz, A. R., Salvadori, M. C., & Cattani, M., (2003). Young modulus measurement of nanostructured palladium thin films. *Technical Proceedings of the 2003 Nanotechnology Conference and Trade Show, 3*, 177–180.

31. Kulkarni, A. V., & Bhushan, B., (1996). Nano/picoindentation measurements on single-crystal aluminum using modified atomic force microscopy. *Materials Letters, 29*(4–6), 221–227.

32. Bamber, M. J., Cooke, K. E., Mann, A. B., & Derby, B., (2001). Accurate determination of Young's modulus and Poisson's ratio of thin films by a combination of acoustic microscopy and nanoindentation. *Thin Solid Films, 398/399*, 299–305.

33. Vilcarromero, J., & Marques, F. C., (2001). Hardness and elastic modulus of carbon–germanium alloys. *Thin Solid Films*, *398/399*, 275–278.
34. Nikolaev, V. I., Shpejzman, V. V., & Smirnov, B. I., (2000). Determination of elastic moduli of GaN epitaxial layers by microindentation technique. *Physics of the Solid State*, *3*, 437–441.
35. Oliver, W. C., & Pharr, G. M., (1992). An improved technique for determining hardness and elastic modulus using load and displacement sensing indentation experiments. *Journal of Materials Research*, *7*(6), 1564–1583.
36. Gogolinskij, K. V., (2015). Sredstva i metody kontrolya geometricheskih parametrov i mekhanicheskih svojstv tverdyh tel s mikro- i nanometrovym prostranstvennym razresheniem [Means and methods for controlling geometric parameters and mechanical properties of solids with micro- and nanometer spatial resolution]. *Dis. dokt. tekhn. nauk.* St.-Petersburg.
37. Useinov, A. S., (2004). Measurement of the a nanoindentation method for measuring the Young's modulus of superhard materials using a nanoscan scanning probe microscope. *Instruments and Experimental Techniques*, *1*, 119–124.
38. Gogolinskij, K. V., Kosakovskaya, Z. Y., Useinov, A. S., & Chaban, I. A., (2004). Measurement of the elastic moduli of dense layers of oriented carbon nanotubes by a scanning force microscope. *Acoustical Physics*, *50*(6), 664–670.
39. Landau, L. D., & Lifshits, E. M., (1987). *Theory of Elasticity.* Moscow: Nauka.
40. Glukhova, O. E., Zhbanov, A. I., & Terentev, O. A., (2002). Theoretical study of the elastic properties of single-walled carbon nanotubes. *Problems of Applied Physics: Intercollegiate Scientific Collection*, *8*, 39–41.
41. Krishnan, A., Dujardin, E., Ebbesen, T. W., Yianilos, P. N., & Treacy, M. M. J., (1998). Young's modulus of single-walled nanotubes. *Physical Review B: Condensed Matter and Materials Physics*, *58*, 14013–14019.
42. Treacy, M. M. J., Ebbesen, T. W., & Gibson, J. M., (1996). Exceptionally high Yong's modulus observed of individual carbon nanotubes. *Nature*, *381*, 678–680.
43. Salvetat, J. P., Briggs, G. A. D., Bonard, J. M., Bacsa, R. R., Kulik, A. J., Stöckli, T., Burnham, N. A., & Forró, L., (1999). Elastic and shear moduli of single-walled carbon nanotube ropes. *Physical Review Letters*, *82*(5), 944–947.
44. Lourie, O., & Wagner, H. D., (1998). Evaluation of Young's modulus of carbon nanotubes by micro-Raman spectroscopy. *Journal of Materials Research*, *13*(9), 2418–2422.
45. Vakhrushev, A. V., & Shushkov, A. A., (2007). *The Method for Determining the Young's Modulus of Elasticity and the Poisson's Ratio of the Material of Micro- and Nanoparticles.* Patent of the Russian Federation No 2297617.
46. Vakhrushev, A. V., Fedotov, A. Y., Vakhrushev, A. A., Shushkov, A. A., & Shushkov, A. V., (2010). Investigation of metal nanoparticles formation mechanisms, determination of mechanical and structural characteristics of nanoobjects and composite materials on their basis. *Chemical Physics and Mezoscopy*, *12*(4), 486–495.
47. Oluwatosin, A. O., (2005). *Synthesis of Porous Films from Nanoparticle Aggregates and Study of Their Processing-Structure-Property Relationships.* Thesis.
48. Bykov, D. L., & Konovalov, D. N., (2000). Osobennosti soprotivleniya vyazkouprugikh materialov pri potere ustoychivosti tonkostennykh konstruktsiy [Features of the resistance of viscoelastic materials in the loss of stability of thin-walled structures].

Fizika protsessov deformatsii i razrusheniya i prognozirovanie mekhanicheskogo povedeniya materialov: trudy 34 Mezhdunarodnogo seminara Aktual'nye problemy prochnosti [The physics of deformation and fracture processes and the prediction of the mechanical behavior of materials: the proceedings of the XXXVI International Workshop Actual Strength Problems]. *Vitebsk,* 428–433.

49. Loboda, O. S., & Krivtsov, A. M. (2005). Vliyanie masshtabnogo faktora na moduli uprugosti trekhmernogo nanokristalla [The effect of scale factor on the elastic moduli of three-dimensional nanocrystal]. *Izvestiya Rossiyskoy akademii nauk. Mekhanika tverdogo tela. [Bulletin of the Russian Academy of Sciences. Solidmechanics], 4,* 27–41.

50. Zavodinskij, V. G., Chibisov, A. N., Gnidenko, A. A., & Alejnikova, M. A. (2005). Teoreticheskoe issledovanie uprugih svojstv malyh nanochastic s razlichnymi tipami mezhatomnyh svyazej [Theoretical study of the elastic properties of small nanoparticles with different types of interatomic bonds]. *Mekhanika kompozicionnyh materialov i konstrukcij [Mechanics of Composite Materials and Structures], 11*(3), 337–346.

51. Bouvier, P., Djurado, E., Lucazeau, G., & Bihan, T., (2000). High-pressure structural evolution of undoped tetragonal nanocrystalline zirconia. *Physical Review B: Condensed Matter and Materials Physics, 62*(13), 8731–8737. https://doi.org/10.1103/PhysRevB.62.8731.

52. Vakhrushev, A. V., & Shushkov, A. A., (2005). Metodika rascheta uprugih parametrov nanoehlementov [Method for calculating the elastic parameters of nanoelements], *Khimicheskaya fizika i mezoskopiya [Chemical Physics and Mesoscopy], 7*(3), 277–285.

53. Vakhrushev, A. V., Shushkov, A. A., & Zykov, S. N., (2013). Sposob opredeleniya modulya uprugosti Yunga materiala mikro- i nanochastic *[The Method for Determining the Young's Modulus of Elasticity of a Material of Micro- and Nanoparticles].* Patent RU 2494038.

CHAPTER 3

Multilevel Computer Simulation: Basic Tool of Nanotechnology Simulations for "Production 4.0" Revolution

A. V. VAKHRUSHEV[1,2]

[1]*Department of Mechanics of Nanostructures, Institute of Mechanics, Udmurt Federal Research Center, Ural Division, Russian Academy of Sciences, Izhevsk, Russia*

[2]*Department of Nanotechnology and Microsystems, Technic Kalashnikov Izhevsk State Technical University, Izhevsk, Russia, E-mail: vakhrushev-a@yandex.ru*

ABSTRACT

The concept of multilevel mathematical modeling with reference to problems of the modern industrial revolution "Production 4.0" is considered. The advantages and disadvantages of accurate mathematical modeling and modeling using artificial intelligence are considered. Examples of the expansion of modeling capabilities using artificial intelligence in space, time, and technological parameters for various nanosystems are given.

3.1 INTRODUCTION

The modern industrial revolution "Production 4.0" requires the translation of all processes preceding the actual receipt of a new product in a digital representation. Forecasts for the development of this stage of production point to the ever-increasing value of computer modeling, the urgency of which will constantly increase. This is because the usual process of creating a new material and introducing it into production usually takes up to 20 years [1]. It is expected that computer modeling will be invested in more financial and intellectual resources. This is especially relevant for one of the main tasks of the industrial revolution "Production 4.0" called "designing materials with controlled properties," which is the basis for the development of nanotechnologies. A full and exact solution to this complex problem is impossible without considering the properties and processes of the formation of materials with controlled properties at the atomic and nanoscale of mathematical description and modeling.

However, a specific feature of the physical processes in nanoscale systems is that the key phenomena determining the behavior of a nanoscale system in real-time at the macroscale take place at small space and time scales. Many experimental and theoretical studies have shown that the properties of a nanoscale system depend not only on the properties of its constituent elements but also on the regularities of the spatial arrangement of the nano-elements in nanosystem and the parameters of the nanoelements interaction.

In this perspective, the multilevel modeling [2–4], which allows describing the formation, evolution, and properties of the nanomaterials in a sufficiently complete and precise manner, should become one of the methods for calculating and modeling modern engineers and technologists. This is explained by the fact that multilevel modeling, being a powerful tool of scientific research, is increasingly becoming a full-ledged stage in obtaining new materials, creating a new technological process and designing a new product at the nanoscale level.

This process was observed in the 1960–1980s of the last century with respect to the finite element method (FEM) and led to the fact that this method is now being applied well and confidently by modern engineers and technologists in the creation of new materials and machines. We can expect that the process of industrial revolution "Production 4.0," the method of multilevel modeling will be adopted as an instrument of engineers and technologists.

Currently, two main approaches to multilevel modeling can be identified. The first method is a method of exact multilevel mathematical modeling, based on an accurate mathematical formulation of the problem. The second method of multilevel mathematical modeling is based on the theory of machine learning and the use of the concept of neural networks and artificial intelligence.

Exact modeling requires a lot of computation time on powerful computers. However, this method has allowed forming, by now, large databases of calculated data on the behavior of nanomaterials and of nanotechnological processes. These data are the base for "learning" the second method, which allows us to significantly speed up the calculation processes on the base of previously obtained results of exact mathematical modeling.

Exact modeling allows us to proceed correctly from the study of atomic and molecular processes by methods of quantum mechanics to the study of processes at the macrolevel by the methods of continuous medium mechanics.

Exact mathematical modeling requires the recording of various mathematical models for the corresponding calculated structural level. This leads to several problems, which are as follows:

- Multiscale nature and connectedness of problems;
- Large number of variables;
- Variation of scales both over space and in time;
- Characteristic times of processes at different scales differ by orders of magnitude;
- Variation of the problem variables at different scales of modeling;
- Matching of boundary conditions at the transition from one modeling scale to another when the problem variables are changed;
- Stochastic behavior of nanoscale systems.

The main problem here is the matching of the boundary conditions of the modeling problems at each space-time scale.

In contrast to this method, machine learning is formed on a single mathematical concept, independent of the structural level of the material under consideration.

The main problems of machine learning mathematical modeling applied to nanosystems include:

- Lack of confidence that this method will lead to the correct result in the field in which there are no exact calculations;
- The training time of the neural network may significantly exceed the time for accurate calculation.

In modern conditions, a successful result is achieved by combining both of the above modeling methods. We give a brief overview of the work on this topic.

The main principles of the multilevel modeling developed for various macro and nanosystems can be found in monographs [2–4].

Consider the work on machine learning in relation to the tasks of designing and researching the properties of new materials.

In paper [5], the Materials Genome Initiative that uses high-throughput computing to uncover the properties of all known inorganic materials is described.

Integrate experimental, computational, and theoretical research tool development and collect digital data available is an actual task of modern material science [6].

Paper [7] introduces a data-driven web-based platform NanoMine for analysis and design of polymer nanocomposite systems under the material genome concept. NanoMine date resource consists of a database, analysis tools, and simulations blocks for processing and investigation of properties and structures of polymer nanocomposites from material selection to macroscopic property prediction.

Databases of material properties of magnetics, thermoelectrics, and photovoltaics functional materials for machine-learning and other data-driven methods for advancing new materials discovery are discussed in paper [8]. Also, in the article, questions for designing new materials that require innovative solutions are formulated.

In paper [9] universal kernel ridge regression-based quantum machine learning (QML) models of electronic properties, trained throughout chemical compound space are presented.

In work [10], the authors propose to use optimized symmetry functions to explore similarities of structures in multicomponent systems in order to yield linear complexity. They combine these symmetry functions with the charge equilibration via neural network technique, a reliable artificial neural network potential for ionic materials.

Paper [11] describes how to speed up the global optimization of molecular structures using machine learning methods. To describe an atomic environment, authors use the symmetry functions proposed by Behler and Parrinello, which ensure rotational and translational invariance.

Machine learning approaches, as productive methods for materials analysis by the emergence of comprehensive databases of materials properties, are described in the paper [12].

In paper [13], the authors developed a new method for a fast, unbiased, and accurate representation of interatomic interactions as a combination of an artificial neural network and a new approach for potential pair reconstruction.

In the article [14], the authors describe the four paradigms of science: empirical, theoretical, computational, and data-driven. They demonstrated how data-driven technique for deciphering processing-structure-property-performance relationships in materials, with illustrative examples of both forward models (property prediction) and inverse models (materials discovery). Also, they illustrate how such analytics can significantly reduce time-to-insight and accelerate cost-effective materials discovery.

The aim of this chapter is to demonstrate the modeling of nanosystems with both above methods. Three problems of modeling are considered and the possibilities of predicting the machine learning method for predicting the behavior of nanosystems in time, when the size of the nanosystem changes and the technological parameters of the nanosystem formation are shown.

3.2 CALCULATION OF THE DEPENDENCE OF THE ELASTIC MODULUS ON THE NANOPARTICLES SIZE

In this section, the computational investigation results on the influence of the size of a nanoparticle under axis tensile forces (Figure 3.2) are presented. The calculations of the equilibrium configuration of nanoparticles show that the particles have the shape close to spherical. Therefore, a sphere was used as an elastic equivalent element. Then, tensile forces are applied to this nanoparticle, as is shown in Figure 3.1, and its load-deformation is investigated.

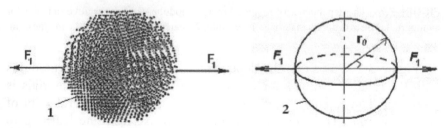

FIGURE 3.1 "Equilibrium" nanoparticle (1) and elastic "equivalent" element (a sphere) (2) stretched by point force.

This task requires a consistent solution of the theory of elasticity and molecular dynamics. A complete mathematical model of the process is presented in [15]. Carrying out the above procedure for nanoparticles of different diameters, we construct the dependence of Young's modulus on the nanoparticle diameter (Figure 3.4). This Figure 3.2 displays the dependence of relative Young's modulus $\bar{E}$ on the radius r for nanoparticles from different materials.

The relation of Young's modulus to its asymptotic value at the maximal nanoparticle diameter was chosen as the relative modulus. To bring the carried-out calculations to a single scale, we divide the nanoparticle radius by its limit radius, for which the relative modulus is 1. In this case, all estimated points are grouped near a generalized curve and can be approximated by a single equation.

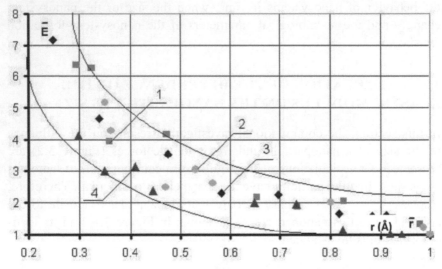

FIGURE 3.2   The dependence of relative Young's modulus $\bar{E}$ on the relative radius $\bar{r}$ for nanoparticles from different materials: 1 – cesium, 2 – calcium, 3 – zinc, and 4 – magnesium.

Figure 3.2 shows that the dependence of relative Young's modulus $\bar{E}$ on the relative radius $\bar{r}$ for nanoparticles from different materials is nonlinear. The calculation of each point on this graph requires a lot of computer time. Therefore, it is advisable to use machine learning to speed up the calculations and expand the predicted range of changes in the modulus of elasticity of the nanoparticles. For this purpose, we construct a graph shown in Figure 3.2 in logarithmic coordinates (Figure 3.3).

In these coordinates, the calculated points are close to the linear graph, so you can use the approximation by a straight line. To determine the coefficients of the equation, we use an artificial neuron with two input signals. The "training" of the neuron and the determination of its weights are carried out by the method of steepest descent (method of rapid descent) [16].

Figure 3.4 shows the results of the "learning" of the neuron. It follows from the graph that the linear dependence well approximates the calculated points and allows one to significantly expand the prediction of Young's modulus of elasticity of the nanoparticle when the spatial parameter (the diameter of the nanoparticle) changes.

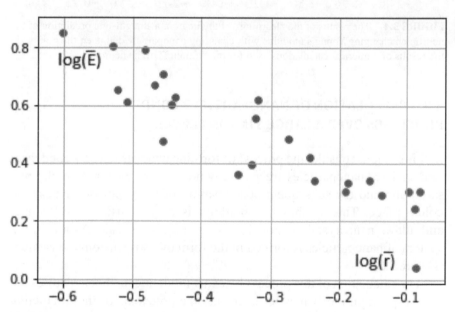

FIGURE 3.3 The dependence of logarithm of relative Young's modulus $\bar{E}$ on the logarithm of relative radius $\bar{r}$ for nanoparticles.

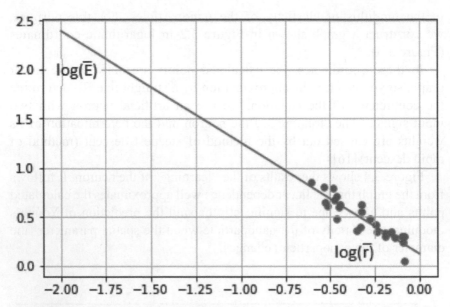

FIGURE 3.4   The results of the "learning" of the neuron: a straight-line prediction of the neuron on changing Young's modulus with changing nanoparticle diameter, the points are the results of numerical calculations used for the "training" of the neuron.

3.3   CALCULATION OF NANOPARTICLE CONDENSATION PROCESSES OVER A LARGE TIME INTERVAL

In this section, consider the process of forming nanoparticles by condensation. Isolated nanoparticles are generally prepared by evaporation, thermal saturation and the subsequent condensation of the vapor on or near the cold surface. This synthesis technology is easy to use, cost-effective, and allows nanocrystalline powders on an industrial scale. As a result, a system of nanoparticles is formed in the form of a suspension in a vacuum (Figure 3.5).

The calculation of this process by exact modeling methods requires the use of quantum mechanics to determine the potentials of the interaction of atoms, molecular dynamics to calculate the process of nanoparticle formation and mesodynamics to simulate the movement and interaction of nanoparticles. A complete mathematical model of the process is presented in [16]. The main problem is that to study the behavior of the nanoparticle system over time, and it is necessary to use a small time integration step

FIGURE 3.5 The computational region in the simulation of nanoaerosol systems formed by condensation.

(between 10^{-15} and 10^{-9} sec) to ensure the required accuracy of the calculations. The real processes of condensation of nanoparticles last a few seconds. Therefore, the solution to this problem by exact methods requires a lot of computer time. Figure 3.6 shows a typical curve of the change over time of the number of nanoparticles in the process of condensation. It can be seen from the figure that at the initial moment of time, the number of particles increases very quickly and then gradually decreases with time. This relationship is highly nonlinear. In logarithmic coordinates, this dependence is linear and therefore, as in the previous section, it is possible to use an artificial neuron with two inputs. Figure 3.7 presents the results of the approximation. The figure shows that an artificial neuron allows approximating the process of condensation of nanoparticles by several orders of magnitude in time.

3.4   SIMULATION OF FORMING NANOCOMPOSITE COATINGS PROCESSES BY ELECTROCHEMICAL DEPOSITION

In this section, we consider the processes of formation of composite nano-coatings in electrocodeposition (ECD) process. Metal matrix composite electrochemical coatings (MMEC) are prepared from the suspensions, representing electrolyte solutions with additives of a certain quantity of

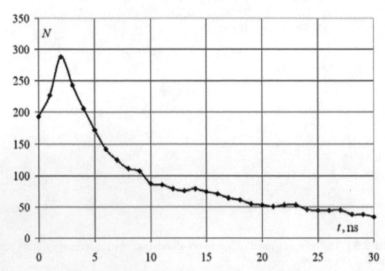

FIGURE 3.6 Changing the number of the nanoparticles formed from the 3-component metal mixture in a calculating cell during the condensation phase.

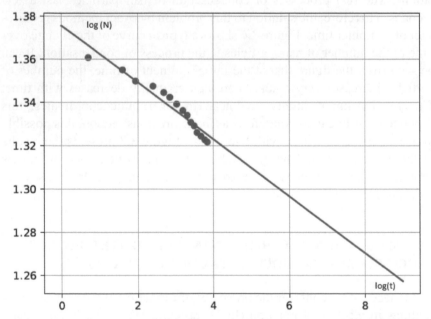

FIGURE 3.7 The results of the "learning" of the neuron: A straight line prediction of the neuron on the condensation process in time, the points are the results of numerical calculations used for the "training" of the neuron.

a superfine powder. The particles are adsorbed onto the cathode surface in combination with metal ions during the ECD process, and the metal matrix composite coating is formed. MMEC consists of galvanic metal (dispersion phase) and particles (dispersed phase).

There are the following steps of the ECD process: (1) the particles in suspension obtain a surface charge; (2) the charged particles and metal ions are transported through the liquid by the application of an electric field (electrophoresis), convection, and diffusion; (3) the particles and metal ions are adsorbed onto the electrode surface; and (4) the particles adhere to the electrode surface through Van Der Waals forces, chemical bonding, or other forces and, simultaneously, adsorbed metal ions are reduced to metal atoms. Metal matrix is encompassed the adsorbed particles, and thus the MMEC is formed.

The ECD process is schematically displayed in Figure 3.8.

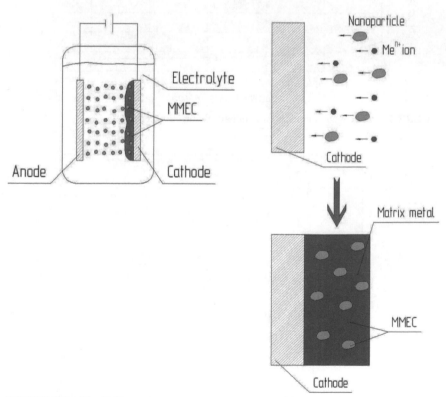

FIGURE 3.8 The ECD process.

Solving this problem requires joint modeling of several problems of hydrodynamic multilevel mathematical modeling and mathematical modeling for mass transfer nanoparticles and electrolyte ions. A complete mathematical model of the process is presented in [17]. The resulting dependencies of weight content on applied current density are depicted in Figures 3.9 and 3.10.

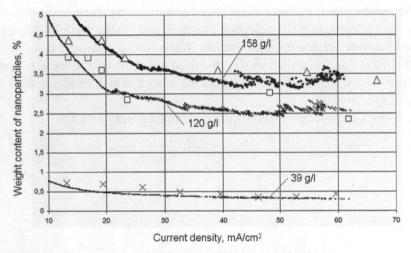

FIGURE 3.9   The weight content of nanoparticle on current density.

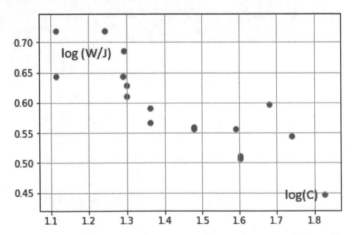

FIGURE 3.10   The dependence of logarithm of weight content of nanoparticles on the logarithm of applied current density: W – weight concentrates of nanoparticles, J – concentration, C – current density.

Figure 3.9 shows the main results of calculations and experimental data [18]. The experimental data are depicted by points. It is obvious that a good agreement with experimental data has been obtained. Therefore, when "learning" artificial neurons, experimental, and calculated data can be combined as educational information (Figure 3.11).

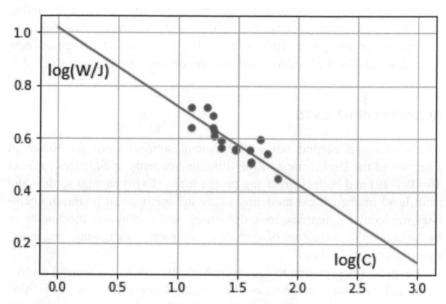

FIGURE 3.11 The results of the "learning" of the neuron: A straight line prediction of the neuron on change of weight content of nanoparticles, when it changes of applied current density, the points are the results of numerical calculations used for the "training" of the neuron.

The experimental and calculated dependences in Figure 3.9 are not linear. In logarithmic coordinates (Figure 3.10), when dividing the experimental and calculated data by the concentration of nanoparticles, the dependence is linear. Therefore, it is possible, as in the previous sections, to use an artificial neuron with two inputs. Figure 3.11 presents the results of the approximation. The figure shows that an artificial neuron allows approximating the dependence on the applied current density with a significant change in the technological parameter (applied current density).

3.5 CONCLUSIONS

In conclusion, it should be noted that machine learning can efficiently combine calculations at separate structural levels using the method of exact multilevel modeling, and also combine the results of experiments and theoretical calculations into a single block of data necessary for training a neural network. High accuracy of exact multilevel mathematical modeling combined with significant acceleration of machine learning modeling is a prerequisite for successfully solving the problems of designing new nanomaterials for the modern industrial revolution "Production 4.0."

ACKNOWLEDGMENTS

The works was carried out with financial support from the Research Program of the Ural Branch of the Russian Academy of Sciences (project 18-10-1-29) and budget financing on the topic "Experimental studies and multilevel mathematical modeling using the methods of quantum chemistry, molecular dynamics, mesodynamics, and continuum mechanics of the processes of formation of surface nanostructured elements and metamaterials based on them."

I express my gratitude to Professor Arzhnikov A.K. for fruitful discussions and valuable comments on the problems and problems of machine learning in relation to the physics of materials. I thank my young colleagues Fedotov A. Yu., PhD, Shushkov, A. A., PhD, and Molchanov E. K. for the calculation and analysis of calculations for exact mathematical modeling.

KEYWORDS

- artificial neuron
- machine learning
- multilevel mathematical modeling
- nanostructures
- nanosystems

REFERENCES

1. Gregory, J. M., & Sean, P., (2016). Paradiso perspective: Materials informatics across the product lifecycle: Selection, manufacturing, and certification. *APL Materials, 4*, 053207.
2. Martin, S. O., (2008). *Computational Multiscale Modeling of Fluids and Solids: Theory and Application.* Berlin–Heidelberg: Springer-Verlag.
3. Weinan, E., (2011). *Principles of Multiscale Modeling.* Cambridge: Cambridge University Press.
4. Vakhrushev, A. V., (2017). *Computational Multiscale Modeling of Multiphase Nanosystems: Theory and Applications.* Apple Academic Press: Waretown, New Jersey, USA.
5. Anubhav, J., Shyue, P. O., Geoffroy, H., Wei, C., William, D. R., Stephen, D., et al., (2013). Persson commentary: The materials project: A materials genome approach to accelerating materials innovation. *APL Materials, 1*, 011002.
6. Pfeif, E. A., & Kroenlein, K., (2016). Perspective: Data infrastructure for high throughput materials discovery. *APL Materials, 4*, 053203.
7. He, Z., Xiaolin, L., Yichi, Z., Linda, S. S., Wei, C., & Catherine, B. L., (2016). Perspective: Nanomine: A material genome approach for polymer nanocomposites analysis and design. *APL Materials, 4*, 053204.
8. Ram, S., & Taylor, D., (2016). Sparks perspective: Interactive material property databases through aggregation of literature data. *APL Materials, 4*, 053206.
9. Felix, A. Faber, A. S., Christensen, B. H., & Anatole, O. V. L., (2018). Alchemical and structural distribution based representation for universal quantum machine learning. *Journal of Chemical Physics, 148*, 241717.
10. Samare, R., Maximilian, A., & Alireza, G. S., (2018). Optimized symmetry functions for machine-learning interatomic potentials of multicomponent systems. *The Journal of Chemical Physics, 149*, 124106.
11. Søren, A. M., Esben, L. K., & Bjørk, H., (2018). Machine learning enhanced global optimization by clustering local environments to enable bundled atomic energies *The Journal of Chemical Physics, 149*, 134104.
12. Eric, G., Cormac, T., Corey, O., Olexandr, I., Fleur, L., Frisco, R., Eva, Z., Jesus, C., Natalio, M., Alexander, T., & Stefano, C., (2017). *AFLOW-ML: A Restful API for Machine-Learning Predictions of Materials Properties.* arXiv:1711.10744v1 [cond-mat.mtrl-sci].
13. Pavel, E. D., Ivan, A. K., & Artem, R. O., (2016). Machine learning scheme for fast extraction of chemically interpretable interatomic potentials. *AIP Advances, 6*, 085318.
14. Ankit, A., & Alok, C., (2016). Perspective: Materials informatics and big data: Realization of the "fourth paradigm" of science in materials science. *AIP Advances, 4*, 053208.
15. Vakhrouchev, A. V., (2009). Modeling of the nanosystems formation by the molecular dynamics, mesodynamics and continuum mechanics methods. *Multidiscipline Modeling in Material and Structures, 5*(2), 99–118.
16. Can, C. A., (2018). *Neural Networks: Evolution.* Moscow. Samizdat.

17. Vakhrushev, A. V., Fedotov, A. Y., & Golubchikov, V. B., (2016). Theoretical bases of modeling of nanostructures formed from the gas phase. *International Journal of Mathematics and Computers in Simulation, 10*, pp. 192–201.

18. Vakhrushev, A. V., & Molchanov, E. K., (2017). "The multilevel modeling of the nanocomposite coating processes by electrocodeposition method," Chapter 14. In: Haghi, A. K., (ed.), *Applied Chemistry and Chemical Engineering* (Vol. 3, pp. 253–312). Waretown, New Jersey, USA: Apple Academic Press.

19. Stojak, J. L., & Talbot, J. B., (2001). Effect of particles on polarization during electrocodeposition using a rotating cylinder electrode. *J. Appl. Electrochem., 31*, 559–564.

CHAPTER 4

On Strengthening of Epoxy-Composites by Filling with Microdispersions of SiC, TiN, and Cement

D. STAROKADOMSKY

Composite Materials Department, Chuiko Institute of Surface Chemistry of the NAS of Ukraine, General Naumov Street, 17, Kyiv-03164, Ukraine, E-mail: km80@ukr.net

ABSTRACT

In this chapter, a new data on the effect of SiC and TiN on the properties of the epoxy polymer is presented. It is established a significant increase in microhardness (in 1.5–2 times), modulus in bending (1.4–1.7 times), compressive strength (for SiC), abrasion resistance and chemical resistance (in nitric acid and acetone/ethyl acetate). At the same time, the thermal stability of the compositions and thermoresistance of their strength properties are substantially increased.

4.1 INTRODUCTION

Epoxy resin is an oligomer with end epoxy groups, usually cured by amines and anhydrides. Her invention of Russian chemists Dianin and Prilezhayev [1] gave beginning to use of epoxy polymers.

Its popularity is due to the ability to turn into a durable, lasting plastic as in industrial as domestic conditions, and most accessible fillers (sand, clay, stone crumbs, chips, etc.) can strengthen him. Its distribution is hampered by the specificity of use (the skills of mixing with the hardener are needed), the toxicity of the hardeners (liquid resin itself is also considered useless), the need for complete drying and degreasing of the bonded surfaces, and other uncomfortable requirements.

As is known, modern ideas about the structure of filled polymers provide for the existence of densified boundary layers near the surface of the filler particles, and a loosened layer of tens of 10–50 mcm thick [2]. The ability of the filler to reinforce is also associated with her formation of chains, frameworks, aggregates, and other filler structures. Schematic of this process is presented in Figure 4.1.

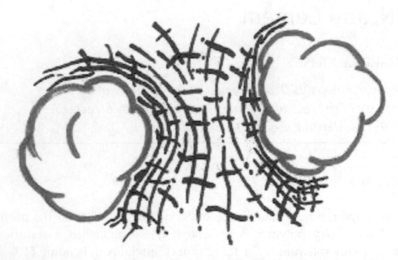

FIGURE 4.1   Author's visualization about a possible dense/loosened polymer chains near filler particles.

The filling of epoxides with cement, carbides, nitrides, and other powders with high specific strength is a known method for obtaining reinforced composites. The filling of powders of carbides and nitrides has long attracted industrialists and researchers [4–11]. The structure of these dispersions suggests a strengthening of polymers after filling.

4.2   EXPERIMENTAL

Strength tests were carried out as follows:

- Compressive strength – on samples-cylinders ($d = 6.5$ mm, height of 11 ± 1 mm).
- Strength in bending (tensile) – on plates $1 \times 5 \times 0.2$ cm^3.

- Microhardness (Rockwell) – with the determination of the force of indentation of the steel hemisphere to a depth of 10–50 mcm.

Abrasion is based on the weight of the friable composite after 60 passes of 20 cm on the P6O emery. Actual resistance is the inverse of the attrition mass X, and the difference in the density of materials must also be taken into account. Therefore, the actual abrasion resistance was determined by the formula $I = \rho/X\rho_0$ (ρ/ρ_0 is the ratio of the densities of the composite and 0% -polymer, X is the weight of erased composite, mg).

Adhesion at tearing – on steel cylinders $d = 5$ cm^2.

Microscopy of the samples was carried out on the GEOL X-ray analyzer.

The thermograms were obtained on the ErdeiPaulich derivatograph in the mode: 100.3 mg sample, 100 mg sensitivity, TG-500, DTG-500, DTA-250, heating rate 10°/min (Figure 4.2).

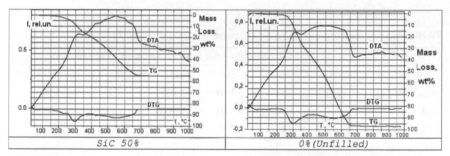

FIGURE 4.2 Thermograms of the composite with silicon carbide (SiC 50%) and unfilled (0%).

Comparison of thermograms does not reveal cardinal changes in the destruction temperatures of samples after filling by SiC (Figure 4.1). Thus, the curves of DTG are almost identical, including the temperature of the minimum (300°C). The temperature of active destruction (a noticeable descent of the TG curve) even decreases after filling (250°C, and 0% polymer –300°C), and 10% – mass loss does not change (300°C).

Only DTA curves (the curve of temperature changes) differ substantially, – the maxima of DTA observed later for 0% polymer (which can speak of an insignificant increase in thermal stability). Differences in mass losses (98% in unfilled and 55% in SiC 50%) are understandable, since carbide does not burn.

The growth of heat resistance after a filling is also manifested in a noticeable increase in fire resistance – in 1.5–2 times (from 1.3 to 2 or 3 seconds, Table 4.1).

TABLE 4.1   The Fire Resistance of Composites by the Time (in Seconds) of Ignition from an Open Fire (the Character of Combustion)

Unfilled	SiC	*SiC/Zement*	TiN
1.3 (self-flames)	2 (self-flames)	3 (weakly flames)	2 (self-flames)

4.3   MICROHARDNESS OF COMPOSITES AT VARIOUS THERMO-TREATMENTS

Thermoresistance of composites is well traced by the example of micro-hardness. The unfilled polymer (Unf) after $t = 55°C$ is very plastic (Table 4.2).

TABLE 4.2   Microhardness of Filled Composites with Different Thermal Modes

	Microhardness at 10–50 mcm of immersion, N				
Soft Term, 55°C 5 hours	10	20	30	40	50 mcm
0% (unfilled)	100	150	230	310	380
SiC	150	210	300	420(X)	550(X)
SiC/Zement	200	250	300 (X)	X	
TiN	200	270	350	440 (T)	550(T)
Hard Term, 250°C 1 hours					
0%	80	170	250	330(X)	X
SiC	100	210	330	430	530
SiC/Zement	100	200	290	390	450(X)
TiN	90	180	300	370	460

Designations: X – fragilely shattered, T – cracked.

Soft heat-treated filled compositions have a higher (of 30–50%) microhardness than unfilled polymer (0-polymer). But they, unlike the O-polymer, are brittle fracture when immersion exceeds 20–30 μm (Table 4.1).

The positive effect of the fillers is noticeable after a hard heat treatment (250°C), when the 0-polymer retains microhardness (Table 4.1), but loses plasticity (the maximal immersion without cracking is only 30–40 μm).

Conversely, a composite with SiC after 250°C acquires a significant plasticity, while maintaining in the parameters of microhardness (Table 4.1). The same we can say about the mixture of SiC/Zement (Table 4.1). That is, composites with SiC give a higher microhardness than the 0% polymer (Table 4.1), while it is much more heat-resistant than the 0%-polymer.

Composite with TiN is initially (after 55°C) more solid than the 0% polymer. But it is much more fragile (tinkering at 40–50 microns of immersion – Table 4.1). After 250°C, its plasticity appreciably increases, but it is accompanied by a drop in microhardness (but it remains higher than for the 0% polymer, Table 4.1). That is, with any heat treatment, the composite with TiN gives a higher microhardness than the 0%-polymer. Moreover, it is also much more heat-resistant (withstands 250°C warm-up) than the 0% polymer.

No less important consequence of filling can be considered reduced shrinkage (Table 4.3).

TABLE 4.3   Shrinkage (mm) of Cylindrical Specimens 12 mm High

0-polymer	SiC	*SiC/Zement*	TiN
1.5	0.8	1	1

4.4   STRENGTH AT BENDING AND COMPRESSION

Epoxy is a "conservative" plastic at bending and compression, and "does not like" dispersed additives. Filling, as a rule, reduces the strength when bending – there was no exception to our case (Table 4.4). Moreover, the bending modulus can still increase significantly after filling – by 1.4–1.7 times. It is most desirable to introduce pure SiC and TiN without cheap additives, since dilution with cement gives the module growth less than 1.4 times (dropping it to the level of composites with pure cement) (Table 4.5).

TABLE 4.4 Bending Strength of Samples

	Strength σ, кg/mm^2 (% to σ for unfilled)	Modulus*10^3, kgs/cm^2
0-polymer (unfilled)	**10.4** *(100%)*	**18.5 (100%)**
SiC	–	32.0 (173%)
SiC/Zement	5.3 (51%)	25.0 (135%)
TiN	6.6 (64%)	28.0 (157%)

TABLE 4.5 Strength (Fracture Load and Modulus) During Compression and Abrasion (ρ and ρ$_o$ are a Density of Filled and Unfilled Composites)

	Abrasion X, mg	Real resistance to abrasion T = ρ/Xρ$_o$	Load of compression (% to H), и modulus E*10^5(% to H)	Load of compression after 250°C
H (0%)	100	0.010 (100%)	430 (100%), E = 15.6 (100%)	340 (100%)
SiC/Zem	120	0.013 (130%)	430 (100%), E = 17.7 (113%)	390 (115%)
SiC	130	0.012 (120%)	470 (109%), E = 16 (103%)	–
TiN	140	0.013 (130%)	370 (86%), E = 14(90%)	350 (103%)

A compression strength X is very conservative characteristic, that is, difficult to increase more than 30–40%. From Table 4.3, we can see that increase of X observes for SiC and Modulus – for SiC/Zement. After 250°C (when O-polymer shows an essential drop of X), the filled composites are more resistant (Tfd.3).

Composite with TiN led to a noticeable drop in both indicators C and X to a level of 70–85% of O-polymer. The nature of the destruction after the addition of TiN (along the Chernov diagonal), here, which makes the destruction elastic (as in the case of SiC and cement).

Some information is given by the diagrams "load-deformation" (Figure 4.2). Unfilled is destroying plastically, with plasticity threshold and the ultimate destruction threshold (Figure 4.2). Destruction for Epoxy+SiC and Epoxy+TiN is elastic, with no signs of plasticity, and always proceeded along the Chernov-Luders diagonal. On the compression diagrams, this is reflected by a change in the curve with a two-humped one on a single-arm. Can see that TiN still leaves some plasticity to the composite (unlike SiC) – in curve we see the similarity of the second threshold of destruction (Figure 4.2). Destruction for Epoxy + SiC/Zement is also elastic – but with a predominance of longitudinal cracks. It refers to the restructuring of the composite structure after filling (Figure 4.3).

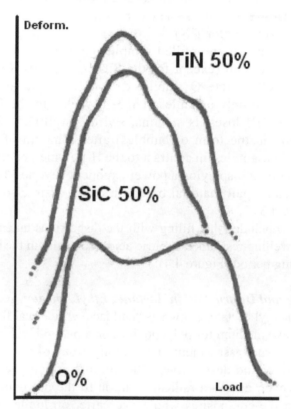

FIGURE 4.3   Compression diagrams.

After filling, the T resistance to abrasion is quite expected (Table 4.3), although it is not as important as in the case of filling with abrasive powders.

4.5   SWELLING AND FIRMNESS IN CORROSIVE ENVIRONMENTS

A   Swelling in 25% HNO₃: Swelling in dilute HNO_3, as is known [3], is not describes by the classical curve. He passes through at least two consecutive stages that are seen from the swelling curves: the first period of active swelling ends in 5–10 days or earlier (for the composition with SiC). After 5–15 days of stabilization (quasi-saturation), the second stage begins, which also ends with a "yield curve on the plateau" on the 30–40th day of soaking (or later for TiN).

It can be seen that an unfilled sample begins to swell more active than all and before all (except for the SiC compound). By the fourth day of his swelling rate Q stabilized at 4%. To this indicator, the filled samples are only suitable after 3–4 weeks. In the future, only a sample with TiN loses its stamina, swiftly swelling (with signs of decomposition in the form of bubbles) after a month of aging. The remaining filled samples, in contrast to the H-polymer, exhibit swelling resistance, without displaying a power exponent beyond 5%. Similarly, the commercial repair material based on epoxy resin also swells (see M1 in Figure 4.3).

It can be concluded that filling with the dispersions taken leads to an increase in swelling resistance in nitric acid, especially in the initial (up to 3 weeks) aging period (Figure 4.4).

B   Swelling and Destruction in Acetone: Ethyl Acetate: Acetone and its mixture with ethyl acetate (known as "nail lacquer removal fluid") are the most aggressive medium for polyepoxides on a par with chloro-carbons.

From the visual assessment of aged only 1 day of samples, one can see how rapidly and destructively the mixture affects the H-polymer. Filling, however, allows a radical solution to the problem of destruction, leaving the composites whole even after prolonged exposure to acetone-ethyl acetate. Only in the case of TiN the sample had insignificant shocks after 1 day of aging (Figure 4.4). In this case, the degree of swelling of Q is markedly reduced. For all samples, we have a similar kind of swelling curves – the maximum at 1 day of exposure is replaced by a decrease in Q (probably due to an increase in the reverse process – leaching).

As you can see, all the fillers (in addition to increased resistance to destruction) make it possible to significantly reduce the activity of swelling in the first hours and days of exposition (Figures 4.5 and 4.6).

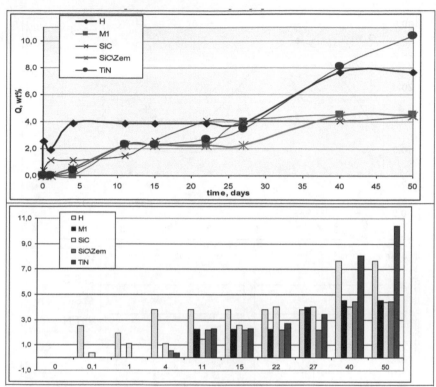

FIGURE 4.4 Curve and swelling histogram of composites in 25% HNO_3. M1 – commercial epoxy-material TM moglice.

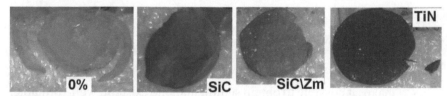

FIGURE 4.5 Type of samples after 1 day of soaking in a mixture of acetone and ethyl acetate. Left to right-Unfilled (destructed); samples: with 50% by weight of SiC; SiC/Cement; TiN (partially destructed).

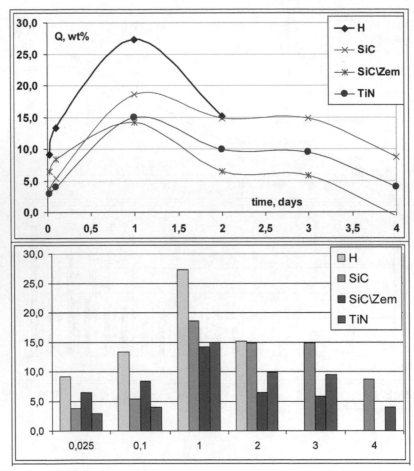

FIGURE 4.6   Curve and swelling histogram in mixture acetone/ethyl acetate.

4.6 CONCLUSIONS

1. The addition of silicon carbide and titanium nitride in epoxy resin makes it possible to obtain sedimentation-resistant, well-formable, and curable shrinkable masses.

2. The DTA method does not reveal significant changes in the intervals of thermo-oxidative degradation after filling. At the same time, the positive effect of fillers on the fire resistance of composites was found.

3. Filling allows to significantly (in 1.3–2 times) increasing the microhardness of composites, but their fragility increases noticeably. A rigid heat treatment (250°C) reveals an additional reserve of hardening by filling, making them plastic (while the unfilled polymer embrittles).
4. The bending strength significantly (1.6–2 times) drops after filling. The modulus of elasticity also increases substantially, especially in the case of silicon carbide filling (1.73 times). In practice, this means obtaining less stress-rattling composites, which, however, crack at lower loads than an unfilled polymer.
5. The compressive strength is significantly dependent on the filler. Substantially falling for the case of TiN, it appreciably increases for SiC, and the SiC/Cement mixture does not change it. The modulus of elasticity does not increase, and the analysis of the diagrams indicates a loss of plasticity after filling (especially for SiC). It is interesting to see the fact of greater thermal resistance (compressive strength to heating) after filling, in particular with the composition SiC/Cement.
6. Filling with the taken dispersions leads to an increase in the resistance to swelling in nitric acid – for all in the initial (up to 3 weeks) holding period, and for SiC/Cement at all stages. A cardinal increase in resistance to acetone and ethyl acetate after filling (especially for SiC/Cement) is noted.
7. The results show little promise as a general reinforcing filler TiN (it only reinforces some characteristics), and the high promise of SiC and its compositions with water-binding agents (e.g., the M400 cement).

KEYWORDS

- **epoxy composite**
- **heat resistance**
- **microhardness**
- **nitric acid**
- **swelling in acetone**

REFERENCES

1. Epoxy: https://en.wikipedia.org/wiki/Epoxy (Accessed on 18 July 2019).
2. Starokadomsky, A. D., (2018). A long century of epoxies. *Science and Life, 1,* 66–71.
3. Lipatov, Y. S., (1991). *Physicochemistry of Filled Polymers* (p. 220). Kiev, USSR, Naukova Dumka (in Russian).
4. Starokadomsky, D. L., (2008). Some features of the swelling of photopolymer composites with different contents of highly dispersed silica. *Plastic Masses* (in Russian) *2,* 33–36.
5. Emelina, O. (2014). *Composite Polymeric Materials,* modified by dispersion fillers, applicable in the construction and repair of equipment. Bulletin of Kazan Technological University, 128–130.
6. Ratna, D., (2005). *Epoxy Composites: Impact Resistance and Flame Retardancy.* RAPRA Review Reports, *16*(5).
7. Abenojar, J., Martínez, M. A., Pantoja, M., Velasco, F., & Del, R. J. C., (2012). Epoxy composite reinforced with nano and micro SiC particles: Curing kinetics and mechanical properties. *The Journal of Adhesion, 88*(4–6), 418–434.
8. Poornima, V. P., Debora, P., Agnieszka, D., Pournami, V., Andrzej, H., Jose, M. K., & Sabu, T., (2015). Mechanical and thermal properties of epoxy/silicon carbide nanofiber composites. *Polymers for Advanced Technologies, 26*(2), 142–146.
9. Shen, D., Zhan, Z., & Liu, J. Y., (2017). Enhanced thermal conductivity of epoxy composites filled with silicon carbide nanowires. *Scientific Reports, 7*(1). doi: 10.1038/s41598-017-02929-0.
10. James, T., (2017). *Fabrication and Characterization of Silicon Carbide Epoxy Composites.* Dissertation PhD, 5, Clemson University, Strengthening of Epoxy-Composites, Filled by SiC and TiN
11. Starokadomsky, D., Sigareova, N., Golovan, S., Tkachenko, A., Moshkovska, N., Kokhtych, L., & Garashenko, I. Scientific Journal, *ScienceRise, 4*(57), 55–59.

CHAPTER 5

Some Innovatives Results-Oriented Scientific Researches That Lead to the Development in the Field of Inorganic Polymer's Science

M. AVALIANI

Tbilisi State University R. Agladze Institute of Inorganic Chemistry and Electrochemistry, Mindeli str. 11 Tbilisi 0186, Georgia, E-mails: marine.avaliani@tsu.ge, avaliani21@hotmail.com

ABSTRACT

The presented work contains information about synthesized by us many crystals and/or powder of inorganic polymeric compounds, concretely diverse new groups of condensed phosphates, during investigations of the systems $M^I_2O-M^{III}_2O_3-P_2O_5-H_2O$ at 400K–850K (where M^I are various monovalent metals, including Ag and M^{III} – some trivalent metals). Numerous – specifically 68 – new formerly unknown double condensed phosphates so-called inorganic polymers have been obtained.

5.1 INTRODUCTION

It is the historical fact that Jöns Jacob Berzelius was not only one of the founders of modern chemistry, but also the first one who has communicated the discovery of the new compound-Condensed Phosphate [1]. Later on, a great number of scientist's researches in the field of chemistry of condensed compounds was published, that are really valuable and very worthwhile [2–7].

Starting from the innovator works of scientists, a lot of condensed compounds, in fact – inorganic polymers was synthesized in the world. A great number of innovative researches in XX Century are really valuable and was appreciated [3–7]. Among varieties of processes of condensation of phosphoric anions, one of them leads to the prearrangement of cyclic, oligomeric or polymeric structures of condensed phosphates. Inorganic polymers-condensed phosphates of polyvalent metals, notably double phosphates containing alkali metals possess a number of rather interesting and treasured properties, which explains prospects of their application. Actually anions are known for $n = 3, 4, 5, 6, 8, 9, 10,$ and 12 [3–10].

5.2 EXPERIMENTAL

5.2.1 MATERIALS

In glassy carbon crucible, there were mixed gallium oxide, or scandium oxide, or indium oxide, orthophosphoric acid (percentage: 85%) and nitrate of silver in various molar ratio.

5.2.2 MEASUREMENT

Several condensed phosphates were prepared during the investigation of poly-component systems $M_2^I O\text{-}M_2^{III} O_3\text{-}P_2O_5\text{-}H_2O$ in the temperature range 400–850K. It is recognized that sufficient stability of polymeric phosphates in this respect makes it possible to identify and categorize them by the method of paper chromatography. This fact permitted scientists to examine the process of formation and the composition of many normal, basic, and/or acid of both simple and double di-, tri-, tetra-, octa, and dodecaphosphates of polyvalent metals. This method, together with the chemical analysis, IR-spectroscopy, thermogravimetry, X-ray diffraction, structural analysis was used by us.

5.3 RESULTS AND DISCUSSION

The aim of our work is obtaining of new inorganic polymeric compounds – condensed phosphates by the ecologically friendly method. The presented data are the results of our studies – synthesis, analysis, examination of the experimental records and their comparison and/or correlation which achievements in the domain of inorganic polymer's chemistry [3–6, 8–10]. Condensed phosphates of polyvalent metals, notably double phosphates containing alkali metals possess a number of rather interesting and valuable, appreciable properties, which explains prospects of their application.

We synthesized many new double condensed oligo- and cyclophosphates, whose general properties we have examined [8–12]; It was executed systematic investigation of systems M^I_2O-$M^{III}_2O_3$-P_2O_5-H_2O at temperature range 400K–850K where M^I – are alkali metals and Ag, M^{III} – Ga, In, Sc, and partially Al. Many compounds were wholly examined, and the structures are determined by X-ray structural techniques [9–13].

Various experiments revealed that by crystallization from melts of polyphosphoric acids were obtained the following double condensed compounds – namely a series of an formerly new class of inorganic polymers: double condensed di- and triphosphates, cyclotetraphosphates, cyclooctaphosphates, cyclododecaphosphates, at the molar ratio P_2O_5:M^I_2O: $M^{III}_2O_3$ = 15:2.5:1,0; 15:5:1.0; 15:7.5:1.0; 15:10:1.0 and 15:3.5:1.5; 15:5:1.5; 15:6:1.5; 15:7.5:1.5; 15:8.5:1.5; 15:12:1.5 [10–13]. In a number of cases, more phosphoric acid was taken during some experiments, (for ex. molar ratio 18:8.5:1.5; 18:10:1.5). Obtained condensed phosphates were detailed examined by thermogravimetric analysis. The thermogravimetric analysis was performed by DTA, the Derivatograph- Q1500-D with a heating rate of 10 degrees/min., in air atmosphere and maximum temperature of 1000K (sometimes to 1400K). Some compounds were also examined by paper chromatography.

Some synthesized phosphates from polyphosphoric acid melts were presented in Table 5.1.

In addition, it was synthesized some polyphosphates $M(PO_3)_3$, e.g., Ga $(PO_3)_3$, In $(PO_3)_3$ in various forms.

More remarkable synthesized condensed phosphates during investigation of the system $M^I O - Sc_2 O_3 - P_2O_5 - H_2 O$ are presented in Table 5.2. The molar ratio for P/MI/MIII was variable: 15/2.5/1; 15/3.5/1; 15/5.0/1; 15/6.0/1; 15/7.5/1; 15/10.0/1; And moreover: 15/2.5/1.5; 15/3.5/1.5; 15/5.0/1.5; 15/6.0/1.5; 15/7.5/1.5; 15/10.0/1.5 as well as: 18/2.5/1.5; 18/3.5/1.5; 18/5.0/1.5; 18/6.0/1.5; 18/7.5/1.5; 18/10.0/1.5 [11–12].

TABLE 5.1 Synthesized Phosphates in Systems $M_2O-M^{III}_2O_3-P_2O_5-H_2O$

$M^I M^{III}(H_2P_2O_7)_2$	$M^I M^{III} P_2O_7$	$M^{III} H_2 P_3 O_{10}$	$M^I M^{III} HP_3 O_{10}$ and/or other complex anions	$M^I_2 M^{III} HP_3 O_{10}$	$M^I M^{III}(PO_3)_4$
$LiGa(H_2P_2O_7)_2$	$LiGaP_2O_7$	$Li_xH_2(2\text{-}X)$	$Li_xH_{2-x}GaP_3O_{10}\cdot(1-1,0)H_2O$	$Li_2GaP_3O_{10}$	$\{LiGa(PO_3)_4\}_x$
$NaGa(H_2P_2O_7)_2$	$NaGaP_2O_7$	$GaH_2P_3O_{10}\cdot H_2O$ form II	$KGaHP_3O_{10}$		$NaGaP_4O_{12}$
$KGa(H_2P_2O_7)_2$	$KGaP_2O_7$		$RbGaHP_3O_{10}$ form I		$K_2Ga_2P_8O_{24}$
$RbGa(H_2P_2O_7)_2$ form I	$RbGaP_2O_7$		$RbGaHP_3O_{10}$ form II		$Rb_2Ga_2P_8O_{24}$
$RbGa(H_2P_2O_7)_2$ form II	$CsGaP_2O_7$		$CsGaHP_3O_{10}$ form I		$Cs_3Ga_3P_{12}O_{36}$
$LiIn(H_2P_2O_7)_2$	$LiInP_2O_7$		$CsGaHP_3O_{10}$ form II		$LiIn(PO_3)_4$
			$CsGaHP_3O_{10}$ form III		$NaInP_4O_{12}$
$LiIn(H_2P_2O_7)_2$	$NaInP_2O_7$		$Cs_2GaH_3(P_2O_7)_2$		
$KIn(H_2P_2O_7)_2$	$KInP_2O_7$		$RbInHP_3O_{10}$		
	$CsInP_2O_7$		$CsInHP_3O_{10}$		

5.3.1 SYNTHESIZED PHOSPHATES IN THE SYSTEMS CONTAINING SCANDIUM AND/OR GALLIUM

Condensed phosphates prepared in poly-component systems, containing Sc and Ga at elevated temperatures are presented in Table 5.2.

In fact, the systems containing Ag-Sc needs to be thoroughly explored at temperature range 620–850K in more depth, which is the objective of our study at the present time. It should be noted that we also received double condensed phosphate – cyclododecaphosphate of Gallium-Silver $Ag_3Ga_3P_{12}O_{36}$. It crystallizes with an impurity second phase at temperature range 605–615K (molar ratio P/Ag/Ga = 18/5/1.3), duration of synthesis is 3–4 days. In our opinion, in the future, it is necessary to carefully select other molar ratios and the continuity of the synthesis to obtain a pure phase. Phase formation in system $M_2^I O\text{-}M_2^{III} O_3\text{-}P_2 O_5\text{-}H_2O$ and the microstructure of synthesized double condensed phosphates are investigated by X-ray diffraction analyses (data summarized in Tables 5.3 and 5.4). The powder diffraction data for cited compounds, intensity data collections are obtained on diffractometer DRON-3M, anodic Cu-Kα radiation, the range $2\theta = 10^0\text{–}60^0$, detector's speed 2^0/min., lattice spacing d_α/n in Angströms Å, and I/I_0 – it is relative intensity (used model/standard data – by American Society for Testing and Materials –ASTM). A detailed comparison with our previously obtained XRD data for similar compounds of Gallium, Indium, and Scandium are also carried out/performed. On the assumption of the fact that combinations of cations (Ag-Ga, Ag-In, and Ag-Sc) for cyclophosphates have not been studied and hence are not given in the file index (typical XRD data), roentgenograms was compared to our standard data models, to similar compounds of Ag-P and our standard data prototypes for similar double condensed phosphates of Gallium, Indium, Scandium with alkali metals [7–13]. We perform that any initial components: K_2CO_3, Sc_2O_3, $AgNO_3$ are already completely irreversibly interreacted during the synthesis process.

TABLE 5.2 Synthesized Condensed Compounds in Poly-Component Systems with Ga and/or Sc

$M^I M^{III} (H_2P_2O_7)_2$	$M^I M^{III} P_2O_7$	$M^I M^{III} HP_3O_{10}$	$M^I_2 M^{III} P_3O_{10}$	$M^I M^{III} (PO_3)_4$	$M^{III} (PO_3)_3$
$LiSc(H_2P_2O_7)_2$	$LiScP_2O_7$	$LiScHP_3O_{10}$	$Li_2ScP_3O_{10}$	$\{LiSc(PO_3)_4\}_x$	$Sc(PO_3)_3$-A
$NaSc(H_2P_2O_7)_2$+ mix phases	$NaScP_2O_7$	$NaScHP_3O_{10}$	$Na_2ScP_3O_{10}$	$Na_3ScP_8O_{23}$ ultra phosphate	$Sc(PO_3)_3$-C
$KSc(H_2P_2O_7)_2$	$KScP_2O_7$	$KScHP_3O_{10}$	$K_2ScP_3O_{10}$	$K_2Sc_2P_8O_{24}$	
$RbSc(H_2P_2O_7)_2$	$RbScP_2O_7$	$RbScHP_3O_{10}$	$Rb_2ScP_3O_{10}$	$Rb_2Sc_2P_8O_{24}$	
$CsSc(H_2P_2O_7)_2$	$CsScP_2O_7$	$CsScHP_3O_{10}$	$Cs_2ScP_3O_{10}$	$Cs_3Sc_3P_{12}O_{36}$	
$Ag(H_2P_2O_7)_2 + AgHSc P_3 O_{10}$ Mix phases	$AgScP_2O_7$	$AgHScP_3O_{10}$		$AgScP_4O_{12}$	
$AgSc(H_2P_2O_7)_2 \cdot H_2O$				$AgGaP_4O_{12}$	

TABLE 5.3 XRD Data for Synthesized Double Condensed Phosphates

$AgScHP_3O_{10}$		ASTM-11-642, $Ag_5P_3O_{10}$		$AgSc(H_2P_2O_7)_2 \cdot 2H_2O$		$NaSc(H_2P_2O_7)_2 \cdot 2H_2O$		$N–Ag–Ga–P$ 5.0 Synthesis at 608K		$Ag–Ga–P$ 7.5 Synthesis at 608K		$NaGaP_4O_{12}$	
d_a/n	I/I_0	d_a/n	I/I_0	d_a/n	I/I_0	d_a/n	I/I_0	d_a/n	I/I_0	d_a/n	I/I_0	d_a/n	I/I_0
–	–	4.81	20	–	–	–	–	4.69	5	4.67	7	4.71	8
4.04	24	–	–	4.04	25	–	–	–	–	4.43	15	4.43	25
–	–	–	–	–	–	4,17	41	–	–	4.20	19	4.27	30
–	–	–	–	–	–	4,00	37	3.94	5	3.94	27	3.94	49
–	–	–	–	–	–	–	–	3.83	100	3.83	80	3.80	15
–	–	3.69	5	–	–	3,67	10	3.63	10	–	–	3.65	8
–	–	3.60	5	–	–	3,59	4	3.56	10	–	–	3.53	5
–	–	–	–	–	–	3,49	25	–	–	–	–	–	–
3.36	72	3.37	10	3.43	33	3,35	29	3.37	31	3.38	38	3.35	16
–	–	3.29	10	–	–	–	–	–	–	–	–	3.29	5
3.20	29	3.21	15	3.32	24	3,24	96	–	–	–	–	3.14	5
–	–	3.15	25	–	–	3,18	8	–	–	3.08	23	3.08	10
3.03	9	3.04	15	3.08	33	–	–	3.03	80	3.03	100	3.03	100
–	–	2.96	50	3.02	30	–	–	–	–	–	18	–	–
–	–	–	–	2.98	100	2,98	100	–	–	–	40	–	–
–	–	–	–	–	–	2,96	19	–	–	–	–	–	–
2.84	100	2.86	95	2.82	55	2,82	52	2.83	19	2.82	15	2.85	20
–	–	2.76	50	–	–	–	–	2.77	20	2.75	13	2.76	10
–	–	2.72	50	2.74	28	2,74	7	–	–	–	–	–	–
–	–	–	–	–	–	2,70	14	2.68	25	2.69	19	–	–
2.64	77	2.60	100	2.64	23	2,62	5	–	–	–	–	2.63	10
2.59	69	–	–	–	–	2,58	4	2.56	15	2.56	15	2.56	3
2.52	14	2.52	30	2.54	32	2,55	7	2.49	16	–	–	2.48	7
2.43	7	2.46	15	–	–	2,47	4	2.41	26	2.40	11	2.40	3
2.37	8	2.38	10	–	–	2,39	7	–	–	–	2.0	–	–

TABLE 5.3 *(Continued)*

AgScHP$_3$O$_{10}$		ASTM-11-642, Ag$_5$P$_3$O$_{10}$		AgSc(H$_2$P$_2$O$_7$)$_2$ · 2H2O		NaSc(H$_2$P$_2$O$_7$)$_2$·2H$_2$O		N–Ag–Ga–P 5.0 Synthesis at 608K		Ag–Ga–P 7.5 Synthesis at 608K		NaGaP$_4$O$_{12}$	
d_a/n	I/I_0	d_a/n	I/I_0	d_a/n	I/I_0	d_a/n	I/I_0	d_a/n	I/I_0	d_a/n	I/I_0	d_a/n	I/I_0
–	–	2.33	5	–	–	2,33	6	2.33	6	–	6	2.32	29
–	–	–	–	–	–	2,30	4	–	–	–	5	–	–
–	–	–	–	2.31	32	–	–	2.25	16	2.25	8	2.20	16
2.21	20	–	–	2.25	18	2,17	12	2.20	12	2.15	8	–	–
2.13	31	2.15	5	2.22	18	2,10	18	–	–	–	–	2.14	5
–	–	2.10	5	–	–	2,07	6	2.07	2	2.06	–	2.13	3
2.03	20	2.00	10	2.03	9	2,04	7	–	5	2.03	–	2.05	3
–	–	1.96	15	–	–	2,02	18	2.00	7	–	–	–	–
1.93	11	1.94	5	–	–	1,95	11	1.92	–	–	–	–	–
1.89	10	1.89	15	1.83	21	–	–	1.87	6	1.89	–	1.87	3
–	–	1.88	10	1.75	11	–	–	1.86	6	–	–	1.85	6
1.76	6	–	–	1.68	22	–	–	1.69	6	1.68	–	–	–
1.67	18	–	–	1.66	28	–	–						
1.64	18	–	–	1.64	33	–	–						
1.62	18	1.59	10	1.57	19	–	–						

TABLE 5.4 XRD Data for Synthesized Double Condensed Phosphates at Relatively High Temperatures

NAg-In-P – 7.5 Synthesis at 608K		NAg-Sc-P – 7.5 Synthesis at 608K		$NaInP_4O_{12}$ Synthesis at 623K		$NaGaP_4O_{12}$ Synthesis at 625K		$Cs_3Ga_3P_{12}O_{36}$ Synthesis at 618K	
d_a/n	I/I_0	d_a/n	I/I_0	d_a/n	I/I_0	d_a/n	I/I_0	d_a/n	I/I_0
4.65	7	4.49	48	4.43	28	4.43	25	5.09	14
4.23	8	4.33	38	4.33	18	4.27	30	–	–
4.04	17	3.94	45	3.95	53	3.94	49	4.00	24
–	–	–	–	3.91	98	–	–	–	–
3.81	13	3.83	50	–	–	3.80	15	3.86	100
3.63	8	–	–	3.66	16	3.65	8	–	–
–	–	3.56	65	–	–	3.53	5	3.59	5
–	–	–	–	3.42	12	–	–	3.50	20
3.39	29	3.36	39	–	–	–	–	–	–
–	–	–	–	3.36	15	3.35	16	–	–
–	–	–	–	3.27	5	3.29	5	–	–
–	–	–	–	3.24	5	–	–	3.25	24
–	–	–	–	3.20	55	–	–	–	–
–	–	–	–	3.15	8	3.14	5	3.14	20
–	–	–	–	–	–	3.08	10	–	–
3.03	100	3.05	100	3.04	100	3.03	100	–	–
–	–	2.89	34	2.93	4	–	–	2.94	40
–	–	2.83	34	2.86	6	2.85	20	2.83	68
2.80	20	2.80	40	–	–	2.76	10	2.76	7
–	–	2.61	16	2.79	22	2.63	10	2.68	14
2.55	15	–	–	–	–	2.56	3	–	–
2.50	15	2.49	20	2.46	8	2.48	7	–	–
–	–	–	–	2.45	15	2.40	3	2.40	6

TABLE 5.4   *(Continued)*

NAg-In-P – 7.5 Synthesis at 608K		NAg-Sc-P – 7.5 Synthesis at 608K		NaInP$_4$O$_{12}$ Synthesis at 623K		NaGaP$_4$O$_{12}$ Synthesis at 625K		Cs$_3$Ga$_3$P$_{12}$O$_{36}$ Synthesis at 618K	
d_a/n	I/I_0	d_a/n	I/I_0	d_a/n	I/I_0	d_a/n	I/I_0	d_a/n	I/I_0
2.27	–	–	–	2.44	5	2.32	29	2,36	40
–	18	–	–	2.32	38	–	–	2.33	10
2.25	22	2.25	25	–	–	–	–	2.27	7
2.23	22	–	–	2.22	4	–	–	–	–
2.20	13	2.20	16	–	–	2.20	16	2.17	4
2.07	6	–	–	2.07	8	2.14	5	2.14	5
–	–	2.06	13	–	–	2.13	3	–	–
1.95	13	2.01	9	–	–	2.05	3	–	–
–	–	1.98	8	1.87	7	1.87	3	–	–
–	–	–	–	1.85	38	1.85	6	–	–
–	–	1.86	5	1.82	5	–	–	–	–
1.78	10	–	–	–	–	–	–	–	–
1.75	10	–	–	1.74	14	–	–	–	–
1.63	11	1.76	8	1.68	5	–	–	–	–
–	–	1.65	13	–	–	–	–	–	–
		1.59	9						

Generally, diverse new groups of condensed phosphates synthesized by us are presented below (see Table 5.5).

TABLE 5.5 Diverse Groups of Synthesized New Condensed Phosphates

Acid triphosphates $M^IM^{III}HP_3O_{10}$ and $M^{III}H_2P_3O_{10}$	Triphosphates $M^I_2M^{III}P_3O_{10}$	Acid diphosphates, hydrated $M^IM^{III}(H_2P_2O_7)_2.2H_2O$
Complex diphosphates $M^I_2M^{III}H_3(P_2O_7)_2$	Acid diphosphates $M^IM^{III}(H_2P_2O_7)_2$	Diphosphates $M^IM^{III}P_2O_7$
Ultra phosphate $M^I_3M^{III}P_8O_{23}$	Long-chain polyphosphates $[M^IM^{III}(PO_3)_4]_x$	$[M^IM^{III}(PO_3)_4]_4$
Cyclooctaphosphates $M^I_2M_2{}^{III}P_8O_{24}$	Cyclododecaphosphates $M^I_3M^{III}_3P_{12}O_{36}$	

5.4 CONCLUSIONS

At relatively low temperatures, it is more probably to produce double acidic phosphates with increasing synthesis temperature double tetraphosphates of Gallium-Silver and Scandium–Silver are formed which are isomorphs among themselves and which are isostructural with the sodium-gallium double condensed tetraphosphates.

By the comparison of the obtained condensed compounds with appropriate phosphates, synthesized by us earlier (in the systems containing Ga, In, and Sc) [7–12], it is possible to conclude that while the radius of trivalent metal decreases, the polyphosphate chain identity period upsurges. This phenomenon is due to the complication of its form-factors.

The less of the correlation/ratio, greater is the possibility of big cycle formation, for example, for obtaining of cycloocta- or cyclododecaphosphates. The optimum achievement for the production of the great cyclic anions is parity of the big cations of monovalent metal versus trivalent metals with small ionic radius.

ACKNOWLEDGMENTS

The support of our colleagues: M. Gvelesiani, D. Dzanashvili, N. Barnovi, E. Shoshiashvili, and Sh. Makhatadze are gratefully acknowledged.

KEYWORDS

- condensed phosphates
- inorganic polymers
- thermogravimetric analysis
- X-ray diffraction analyses
- trivalent metals

REFERENCES

1. Berzelius, J., (1816). Studies on the composition of phosphoric acid and its salts. *Annals of Physics, v. 54, Joh. Ambr. Barth, Leipzig*, 31–52.
2. Averbouch-Pouchot, M. T., & Durif, A., (1996). *Topics in Phosphate Chemistry* (p. 404). World Scientific.
3. Durif, A., (2014). *Crystal Chemistry of Condensed Phosphates* (p. 408). Springer Science & Business Media.
4. Tananaev, I. V., Grunze, X., & Chudinova, N. N., (1984). Prior directions and results in the domain of condensed phosphates' chemistry. *J. Inorgan. Mater., 20*(6), 887.
5. Murashova, E. V., & Chudinova, N. N., (2001). Double condensed phosphates of cesium-indium. *J. Inorgan. Mater., 37*(12), 1521.
6. Zanello, P., (2012). *Chains, Clusters Inclusion Compounds, Paramagnetic Labels, and Organic Rings.* University of Milan, Elsevier.
7. Durif, A., (2016). *The Development of Cyclophosphate Crystal Chemistry* (p. 12). LEDSS, Docslide US.
8. Avaliani, M., Purtseladze, B., Gvelesiani, M., & Chagelishvili, R., (2011). *The Characterization of the New Group of Inorganic Polymers-Condensed Phosphates* (p. 55). Tbilisi State University, Intern. Conf. GEOHET, book of abstracts.
9. Avaliani, M., Gvelesiani, M., Barnovi, N., Purtseladze, B., & Dzanashvili, D., (2016). Investigations of poly-component systems in aims for Synthesis of a new group of inorganic polymers-condensed phosphates. *Proceed of the Georgian Academy of Sciences, Chem.ser 42*(4), 308.
10. Avaliani, M., (2015). *General Overview of Synthesis and Properties of a New Group of Inorganic Polymers – Double Condensed Phosphates* (p. 240). Intern. Conf. ICAMT.
11. Avaliani, M., Shapakidze, E, Barnovi, N., Gvelesiani, M., & Dzanashvili, D., (2017). About new inorganic polymers-double condensed phosphates of silver and trivalent metals. *J. Chem. Chem. Eng. USA, 11*, 60.
12. Avaliani, M., (2016). New inorganic polymers-condensed phosphates obtained in multi-component systems from solution-melts of polyphosphoric acids. *J. Nano Studies, 13*(2), 135.
13. Avaliani, M. A., (1982). *Synthesis and Characterization of Gallium Indium Condensed Phosphates* (p. 185). Extended Abstract of Cand. Sci. Dissertation.

CHAPTER 6

Nanocomposites for Environmental Protection: Technological Vision and the Next-Generation Environmental Engineering Technique

SUKANCHAN PALIT

*43, Judges Bagan, Post-Office-Haridevpur, Kolkata – 700082,
India, Tel.: 0091-8958728093,
E-mails: sukanchan68@gmail.com, sukanchan92@gmail.com*

ABSTRACT

Science and technology globally are in the avenues of vast scientific regeneration. Nanotechnology (NT) and environmental protection are in the similar vein and similar vision surpassing vast and versatile scientific boundaries. Environmental NT is a frontier area of science and engineering today. Environmental pollution control, drinking water treatment, and industrial wastewater treatment are the utmost needs of human civilization today. Environmental NT addresses these aspects of scientific endeavor and opens a newer school of thoughts and scientific instinct in the field of science and engineering. Mankind's immense scientific provenance, man's vast scientific grit and determination, and the futuristic vision of NT will all lead a long and visionary way in the true realization of environmental remediation and groundwater remediation. Environmental NT today is one of the futuristic avenues of environmental engineering research pursuit. Nanocomposites are in the path of new scientific regeneration and newer scientific redeeming. Technological validation, scientific motivation, and deep scientific vision of environmental NT will surely unfold the vast scientific truth of environmental protection and water pollution control. In this treatise, the author elucidates on the recent scientific endeavor and the scientific emancipation in the field of global NT applications. The author profoundly depicts the success of science, the visionary avenues of engineering science and the needs of human society in the true realization of environmental remediation, groundwater remediation, nanotechnology and nanocomposites. The success of engineering science and technology of green NT, green engineering and green chemistry are elaborated in minute details in this treatise.

6.1   INTRODUCTION

The world of environmental engineering science is today witnessing immense scientific challenges and deep scientific forbearance. Frequent environmental disasters, global climate change, and the immense scientific potential of nanotechnology (NT) and environmental NT will all be the definite forerunners towards a newer era in science and engineering globally. Technology and engineering science need to be re-envisioned and re-envisaged with the passage of scientific history, scientific vision, and time. Stringent environmental regulations globally have urged science and technology of environmental protection to take concrete steps and ensure vast innovations. Provision of basic human needs today include energy, water, food, education, and shelter. Sustainability, whether it is energy, environmental, social, or economic, is the needs of human society today. In the similar vein, the provision of clean drinking water and zero-discharge industrial processes are the pillars of scientific endeavor today. The challenges and the vision of environmental protection and groundwater remediation are dealt with an immense scientific conscience in this chapter. Developing and developed nations around the world are faced with the world's largest environmental disaster – the arsenic groundwater poisoning, and thus, the need for environmental NT in groundwater remediation. In this treatise, the author deeply elucidates on the immediate needs of energy and environmental sustainability in the furtherance of science and technology globally. Today the success of science and technology in the global scenario depends on proper implementation and the larger emancipation of energy and environmental sustainability. Sustainable development is the coined word of today's human civilization. Technology and engineering has practically no answers to the many research questions of water purification and industrial wastewater treatment. Groundwater remediation is a valid research question towards the progress of science and technology. Thus, the need for NT applications and greater environmental NT emancipation.

6.2 THE AIM AND OBJECTIVE OF THIS TREATISE

Human civilization and human scientific progress today stands in the midst of immense scientific vision and sagacity. Global environmental restrictions and stringent environmental regulations have urged scientists and engineers to gear forward towards newer environmental engineering innovations and techniques. Thus this is the primary aim and objective of this treatise. The challenges and the vision of groundwater remediation and environmental remediation are vast and versatile. The author depicts profoundly the scientific success, the vast scientific ingenuity, and the scientific provenance of the field of environmental NT. The author in this treatise pointedly focuses on the vast applications of environmental NT in diverse areas of human scientific endeavor. Nanomaterials (NMs) and engineered NMs are the next generation materials and today are overcoming vast and versatile scientific frontiers. Nano-engineering is a field which has immense scientific potential and is replete with ingenuity and scientific vision. Thus in a similar vision, the author also targets the NT research initiatives in developed and developing nations around the world. The state of environmental engineering and environmental remediation is immensely dismal. Science has practically no answers to heavy metal groundwater poisoning and global poisoning. The pillar of this treatise is to unravel the deep scientific intricacies, the success and the vision of groundwater remediation techniques and NT applications. Human factor engineering and systems engineering approach and the application of technology management in NT applications in environmental protection are the other pillars of this chapter. A deep scientific vision and vast scientific redeeming in the field of nanocomposites are the other pillar of this article. Human scientific rigor and academic motivation will then usher in a new era in environmental remediation.

6.3 THE VAST SCIENTIFIC DOCTRINE OF NANO-SCIENCE AND NANOTECHNOLOGY (NT)

The vast scientific doctrine of nano-science and NT is far-reaching and are in the path of newer scientific regeneration. NT today is integrated with diverse branches of science and engineering, which includes environmental protection science. Environmental sustainability in the similar vein is the imminent need of human society today. Today science and engineering globally are highly advanced and needs to be re-envisioned and re-envisaged with the progress of scientific and academic rigor. Today is the age of vast scientific understanding in the field of space technology, nuclear science, and renewable energy. Human mankind's immense scientific knowledge, man's immense engineering prowess, and the technological profundity are the torchbearers towards a newer era in the field of environmental protection and NT. In a similar vision, renewable energy and NT needs to be integrated with each other for the future emancipation of science and engineering. Biogas energy, solar energy, and wind energy are the needs of humanity today. Thus the modern science and present-day human civilization are highly dependent of immense scientific strides in NT and sustainability.

NT is the intrinsic manipulation of matter on an atomic, molecular, and vastly supramolecular scale. The earliest description of NT referred towards a technological vision of precisely manipulating atoms and molecules for fabrication of macroscale products and now referred to as molecular NT. Another definition of NT is the manipulation of matter with at least one dimension sized from 1 to 100 nanometers. NT is the engineering of functional systems at the molecular scale. This encompasses both current work and scientific concepts at an advanced level. In its original sense, NT refers to the targeted ability to construct items from the bottom up, using tools being developed today to make complete, and high-performance products.

6.3.1　THE VISION OF THE APPLICATION OF NANOCOMPOSITES IN ENVIRONMENTAL NANOTECHNOLOGY AND ENVIRONMENTAL POLLUTION CONTROL

Nanocomposites are today the needs of physical science and material science. Composite science and nanotechnology are the immediate visionary domains of research pursuit. Environmental nanotechnology and industrial pollution control are two undivided areas of science and engineering. Scientific vision and vast scientific transcendence are the needs of environmental nanotechnology and nanocomposites research today. The challenges, the targets and the scientific revelation will surely open new future recommendations and newer futuristic thoughts in the field of nanocomposites. A brief discussion in the field of nanocomposites application will widen the area of environmental remediation. Composite science, material science and polymer science will surely open up new vistas of research pursuit in years to come. The scientific travails and the engineering emancipation will clearly widen humanity's research pursuit. The author deeply pursues these areas.

6.4 ENVIRONMENTAL REMEDIATION AND THE SUCCESS OF SCIENCE

Environmental remediation and the success of science and engineering are the needs of human civilization and human scientific progress today. Groundwater remediation and decontamination of drinking water and groundwater are the utmost needs of humanity today. Today the success of science and engineering depends on the application of sustainability science to human society. Energy and environmental sustainability are the pillars of the holistic development of human society today. Environmental disasters, the loss of ecological biodiversity, and the grave concerns of global warming have propelled engineers and scientists to take serious steps in mitigation of this global crisis.

6.5 ENVIRONMENTAL NANOTECHNOLOGY (NT) AND THE VISION FOR THE FUTURE

The integration of environmental engineering and NT is the need of human society today. Environmental NT emancipation is the primary need of environmental remediation today. The challenges and the vision of NT today globally have no boundaries. The author in this paper rigorously points out towards the success of environmental sustainability, environmental engineering, and NT in the furtherance of science and technology. The creation and realization of the science of NT is human civilization's greatest marvel. Scientific splendor, deep scientific inquiry, and scientific divinity are the utmost needs of research pursuit and deep emancipation today. The vision for the future of NT is immense, thought-provoking, and scientifically sounds today. The author repeatedly points towards the imminent needs of environmental NT in the success of human scientific rigor and vast environmental engineering rigor.

NT is today used in several applications to improve the environment. This is the basic fundamental of environmental NT. This includes cleaning up existing pollution, improving manufacturing methods to reduce the generation of new pollution, and enhancing the world of renewable energy. Renewable energy and NT are the needs of human civilization and human scientific endeavor today. The applications of NT to tackle environmental issues are:

- Generating less pollution in the manufacture of advanced materials.
- Producing solar cells that generate electricity at a lower cost.
- Increasing the electricity generated by windmills.
- Cleaning up organic pollutants in wastewater with the help of NT and environmental remediation.
- Cleaning up oil spills.
- Clearing volatile organic compounds from air.
- Reducing the cost of fuel cells and enhancing energy sustainability.
- Storing hydrogen for fuel cell cars.

Technology and engineering science of energy and environmental sustainability are gearing forward towards a newer dawn in human scientific progress. Today NT and environmental remediation are the needs of human civilization. In this chapter, the author depicts profoundly the utmost needs of energy and environmental sustainability in the furtherance of science and technology globally.

6.6   SCIENTIFIC INQUIRY, SCIENTIFIC INGENUITY AND THE ROAD TO SCIENTIFIC WISDOM IN THE DOMAIN OF NANOTECHNOLOGY (NT)

Technology and engineering science in the global scenario are in the path of newer scientific regeneration. The scientific inquiry, the scientific ingenuity, and the scientific divinity in the field of NT today has no parallels. Water crisis and global warming are veritably destroying the scientific firmament. Today basic and applied sciences are changing the face of human scientific progress. Energy engineering, renewable energy, and environmental engineering are the utmost needs of human civilization. The road to scientific wisdom is vast and versatile as science and engineering move forward towards a newer visionary era. In this treatise, the author deeply stresses on the application of NT and environmental NT to the human society. Provision of pure drinking water is in the path of newer scientific validation and deep futuristic vision. This treatise is a veritable eye-opener towards the visionary scientific intricacies in the field of NT and nano-engineering. Science and engineering of NT are vastly ebullient, and thought-provoking as civilization moves forward.

6.7 SIGNIFICANT SCIENTIFIC RESEARCH PURSUIT IN THE FIELD OF NANOTECHNOLOGY (NT)

NT and nano-engineering are today in the wide path of scientific divination, provenance, and vision. Research pursuit needs to be revamped and re-organized as science and engineering move forward. A brief discussion on the field of NMs and engineered NMs are done with vast scientific far-sightedness.

Arfin et al., [1] discussed with cogent and lucid insight engineered NMs for industrial applications. Today the world is witnessing various environmental problems such as pollution of a different kind, global warming and climate changes, greenhouse gas effects, and the monstrous problem of heavy metal drinking water contamination [1]. Thus technology needs to be revamped and re-envisioned [1]. The authors discussed the various applications of NMs such as carbon nanotubes, fullerenes, graphene (GR), zinc oxide, titanium dioxide, dendrimers, and carbon nanodots. Iron-containing nanomaterial, polystyrene, chitosan, gelatin, polyaniline, bentonite, and cellulose are the other areas of scientific research pursuit and the scientific rigor of this chapter [1]. The manufacturing areas of NMs and its physical and chemical methodologies are the other salient features of this treatise. Human scientific endeavor in NT and nano-engineering are in the path of newer scientific regeneration [1]. This chapter opens up newer thoughts and newer scientific vision in NT applications in human society [1].

Sen et al., [2] discussed with immense scientific conscience biotechnology and its applications in environmental remediation. This chapter explores biotechnological modes, updated in terms of environmental remediation [2]. Modern biotechnology today stands in the midst of deep scientific vision and deep scientific contemplation. This chapter discusses the removal of arsenic in aquifers, treatment of tannery effluents, and the application of reverse osmosis in removing recalcitrant pollutants. An integrated approach combining mineral processing techniques, NT, nano-engineering, and biotechnology are the recommendations of this treatise [2]. Bioremediation, groundwater bioremediation, oil bioremediation, and biosorption of harmful pollutants are the other major pillars of this well-researched treatise [2]. Removal of arsenic from groundwater source stands as a major impediment to the success of science and engineering in the developing and developed nations around the world. A deep investigation on the success and ingenuity of bioremediation are dealt with immense scientific revelation in this paper [2].

Rajan, [3] discussed and elucidated with immense vision and introspection NT in groundwater remediation. Nanoparticles (NPs) are the key components that have promised many benefits through their nano-enabled applications in varied sectors. This well-researched review summarizes the importance of NMs such as zero-valent iron and carbon nanotubes in environmental cleanup like groundwater remediation for drinking and reuses [3]. Technological vision, the vast scientific profundity, and the scientific ingenuity of groundwater remediation and NT applications are dealt with in minute details in this paper [3]. Nanoremediation is a veritable pillar of this treatise. In recent years, nanoremediation has become the primary focus of research and development. This research pursuit pointedly focuses on the scientific success, the vast scientific divination, and the profundity of engineering of nano-remediation [3].

Environmental engineering science, chemical process engineering and NT need to be integrated today as civilization moves forward. Water purification and novel separation processes are the two opposite sides of the visionary coin. Today, arsenic, and heavy metal drinking water poisoning are the bane of human civilization and academic rigor. The immediate necessity of science and engineering are the greater emancipation and the greater realization of nano-science and NT. Today the fruit of human civilization is NT. The author rigorously points towards the vision and the targets of NP applications in environmental protection and chemical process engineering.

6.8 SIGNIFICANT SCIENTIFIC ENDEAVOR IN THE FIELD OF ENVIRONMENTAL REMEDIATION AND ENVIRONMENTAL NANOTECHNOLOGY (NT)

Scientific research pursuit in the field of environmental remediation today stands in the midst of vision and vast scientific contemplation. NT or environmental NT is today integrated with diverse areas of science and engineering globally. The vast candor of science and engineering and technological validation are the torchbearers towards a newer era in the field of environmental protection and environmental NT. Science and engineering are today on the verge of newer scientific rejuvenation with vast strides in the field of nuclear science and space technology. In a similar vein, renewable energy and NT are the next-generation scientific frontiers. In this section, the author deeply comprehends the scientific success, the deep scientific ingenuity, and the vast world of scientific inquiry in the field of environmental NT.

Monteiro-Riviere et al., [4] deeply discussed with lucid and cogent insight toxicological impacts of NMs. NT and nanoengineering have tremendously stimulated fresh interest in the role of particle size in determining toxicity [4]. Health effects due to the application of NMs stands as a major pivot towards the emancipation of NT and nanoengineering. NPs may be more toxic than larger particles of the same substance because of their larger surface area, a higher ratio of particle number to mass, enhanced chemical reactivity and vast potential for easier penetration of cells [4]. Due to the widespread use of NMs, it is essential that scientists understand and explore the biological effects of exposure for their medical, occupational health, and environmental effects. Since their first discovery, NMs have been proposed for vast use in many biological applications, although little is known of their toxicity, potential mutagenic effects, or overall risk to human health [4]. New NMs should be thoroughly investigated for occupational safety during manufacture, exposure scenarios likely to be encountered by the consumer (e.g., commercial products, medicines, cosmetics) and post-use release and migration to the environment [4]. This chapter deeply investigates the known toxicity of several classes of NMs. The areas covered in this chapter are fullerenes, single-walled carbon nanotubes, multi-walled carbon nanotubes, complications in screening assays using carbon-based NMs, titanium dioxides, iron oxides, cerium dioxides, copper NPs, gold NPs, quantum dots (QDs), exposure, and risk assessment, and the vast area of environmental impact [4]. Deep scientific ardor, the futuristic vision of environmental protection and NT will surely uncover the scientific intricacies and the scientific barriers of engineering and technology of NMs [4]. Fullerenes or buckyballs are molecular structures containing 60 carbon atoms. These spheres are made up of carbon, and each carbon is bonded to three neighboring carbons of about one nanometer in diameter. Three scientists first discovered fullerenes in 1985 while studying "clusters-aggregates of atoms" where they vaporized graphite with a laser beam. [4] In 1996, the Nobel Prize was awarded to Curl, Kroto, and Smalley for their path-breaking and remarkable discovery [4]. That was a ground-breaking discovery in the field of science and engineering which surpassed many scientific frontiers. The discovery of fullerenes paved the way for the discovery of carbon nanotubes, single-walled carbon nanotubes, and multiwalled carbon nanotubes [4]. Technological vision, the deep scientific understanding and discernment and the vast world of scientific validation will surely open up newer thoughts and newer visionary avenues in the field of both environmental NT and NT [4].

Mansoori et al., [5] discussed with immense scientific conscience and deep academic brilliance environmental applications of NT. NT is a wonder of science today and revolutionizing diverse branches of science and engineering. [5] As discussed earlier, today, environmental protection science, industrial water resource management, and human factor engineering needs to be integrated with each other. In a similar vein, NT needs to be connected with environmental engineering science [5]. Today NT has immense potential in the development of newer and innovative products [5]. They also have the capability to reformulate new materials and chemicals and reduced harm to the environment as well as targeting environmental remediation. This chapter gives a comprehensive review on the present research endeavor on environmental remediation by NT. In the beginning, the essential aspects of environmental problems are broadly reviewed, and then the application of NT to the compounds, which serves as environmental cleaning are deeply elucidated. Various environmental treatments using nanostructured materials are discussed in deep details [5]. The broad categories of NPs studied are titanium dioxide, iron, bimetallics, catalytic particles, clays, carbon nanotubes, fullerenes, dendrimers, and magnetic NMs [5]. The vision of industrial applications of NPs is today vast and versatile. This paper is a veritable eye-opener towards the utmost needs of NMs for the progress of human civilization. NMs and their environmental applications are the pivots of this scientific research pursuit.

Report of the National NT Initiative Workshop, USA [6] deeply elucidated with scientific conscience NMs and the environment and instrumentation, metrology, and analytical methods. The National NT Initiative is the Federal NT Research and Development program established in 2000 to coordinate and envision NT research, development, and deployment [6]. The world of science and technology today has an utmost need that is NT. The domain of NT and nano-engineering thus needs to be re-envisioned and re-organized at every step scientific and academic rigor globally. NT holds the immense promise of exciting new solutions to crucial scientific, industrial, and commercial challenges through the engineering of NMs [6]. Engineered NMs are already being used in the global market place and in the global scientific scenario. New and novel NMs are frequently used for commercialization [6]. Technology, engineering science, medical science, and systems engineering thus needs to be re-envisaged and revamped with the passage of scientific history and time. Thus NMs and the environment research needs are:

- Understand the effects of engineered NMs in species and individuals.
- Understand the environmental exposures through the principal sources of exposure.
- Determine the factors for environmental transport of NMs.
- Understand the transformation of NMs [6].
- A larger scientific emancipation for evaluation of abiotic and ecosystem-wide effects [6].

A deep analysis of the environmental effect of NMs and engineered NMs is done with vast scientific vision and conscience. A redefinition of the transport processes and the colloidal science of engineered NMs are deeply done in the deliberations of this workshop. Science and engineering thus need to be revamped as regards application of NMs and engineered NMs to humanity.

Palit et al., [7] discussed and elucidated with vast scientific far-sightedness engineered nanomaterial for industrial use. Engineered NMs are the next generation smart materials and have diverse applications. The authors in this chapter deeply discussed the different types of NMs and its wide applications. The different Engineered Nano Materials are carbon black, carbon nanotubes, dendrimers, fullerenes, GR sheets, metal oxides, metals, nanoarrays, nanocrystals, nanowires, and QDs [7]. Scientific research pursuits in Engineered NMs are today opening new avenues of scientific innovation and scientific instinct in the field of NT [7]. The authors also deeply discussed environmental sustainability and its wide vision for the future. Today perspectives of science and engineering are vast and versatile. In this treatise, the authors profoundly depict the vast interface of the application of Engineered NMs and the true realization of NT [7].

Palit, [8] discussed with vast scientific vision and foresight engineered NMs in the environment industry. The world of environmental remediation and environmental engineering science are today in the midst of deep scientific introspection and scientific and engineering ingenuity [8]. The challenges and the academic rigor of NMs are today highly advanced and far-reaching. Engineered NMs are materials created by manipulation of matter at the nanoscale to produce new materials, structures, and intelligent devices [8]. The author in this chapter deeply discussed NMs and environmental engineering science. This well-researched treatise also discussed recent scientific research pursuit in the field of NT and a deeper emancipation of nano-engineering [8]. The other areas of research pursuit in this treatise are the recent endeavor in the field of NMs. The larger scientific vision behind global water challenges are delineated and deeply elucidated in minute details. Groundwater remediation and the futuristic vision of human scientific endeavor are the other highlights of this treatise. Environmental engineering science and environmental protection today stands in the midst of deep scientific comprehension and scientific contemplation. There are today no answers to the intricate questions of groundwater and drinking water contamination. Application of nanoadsorbents, nanofiltration, and nanostructured membranes in groundwater remediation are the areas of scientific research pursuit which needs to be envisioned and re-organized [8]. This chapter is a veritable eye-opener towards a newer era in the field of NMs applications in environmental protection [8].

Pathakoti et al., [9] discussed with deep scientific vision NT applications for environmental industry. The authors discussed in details water and wastewater treatment, nanoadsorbents, nanocatalysts, nanomembranes, and NPs for remediation [9]. The other areas of research endeavor are energy sustainability and energy sources. Energy storage and energy-saving and its technologies are delineated in deep details in this chapter [9]. The vast area of environmental sensing and its various devices are vastly deliberated with immense scientific vision. The possible environmental, social, and ethical implications should be considered in deep details when NMs applications to human scientific rigor are taken into account. This chapter is a well-researched treatise on the environmental implications of NMs applications [9].

Human scientific endeavor and vast scientific and academic rigor in the field of NMs and engineered NMs are vast, versatile, and groundbreaking. Science and technology in this decade are today surpassing vast frontiers. Human scientific imagination is at its best today as nano-technology and nano-engineering change the vast scientific fabric. In this chapter, the author repeatedly pronounces the scientific needs and the scientific ingenuity in the field of NT and environmental protection. The challenges and the surmounting vision are deeply related in this paper.

6.9 ENVIRONMENTAL PROTECTION AND THE SCIENTIFIC WISDOM TOWARDS THE FUTURE

Environmental protection and water purification are the much-needed areas of scientific research pursuit today. Technological vision and deep ardor, the validation of science and engineering and the vast scientific wisdom of environmental engineering science will all lead a long and visionary way in the true emancipation of energy and environmental sustainability. The imminent need of the hour in the global scientific scenario is energy and environmental sustainability. Nanoengineering, environmental NT, and environmental systems management are the equal needs of the hour globally today. Systems engineering approach in solving intricate engineering problems are the needs of engineering, technology, and science today. Water purification, drinking water treatment, and industrial wastewater treatment needs to be re-envisioned and re-envisaged with the passage of scientific history and time. In this treatise, the author deeply comprehends the world of challenges and the vision behind environmental protection and integrated water resource management. The scientific wisdom and the deep scientific inquiry will surely unfurl the difficulties and barriers of environmental remediation and environmental sustainability. In a similar way, the challenges of the application of environmental sustainability to humanity will open new vistas of scientific endeavor globally [10, 11].

6.10   HEAVY METAL GROUNDWATER REMEDIATION AND THE SUCCESS OF ENGINEERING AND TECHNOLOGY

Heavy metal groundwater remediation is the imminent need of science and technology today. The success of engineering and technology is today retrogressive as civilization surges forward. Human civilization and human scientific progress today stands in the crucial crossroads of vision, scientific might, and vast scientific contemplation. The extent of heavy metal and arsenic drinking water poisoning in Bangladesh and India is immense and vastly thought-provoking. The pillars of engineering and science such as chemical process engineering, NT, sustainability science, and human factor engineering are of utmost importance as human civilization moves forward. Today the success of environmental engineering depends on concerted scientific efforts in the field of technology management and engineering systems management. Arsenic drinking water poisoning is destroying the vast scientific firmament of numerous developing and developed nations around the world. Technological impasse, deep scientific inquiry and scientific forbearance in such a crucial juncture will lead a long and visionary way in unraveling the truth and fortitude of environmental engineering science and chemical process engineering. Heavy metal groundwater remediation needs to be done with immense vision, and scientific rigor as science and engineering overcomes one frontier over another.

6.11   ARSENIC GROUNDWATER CONTAMINATION, THE MITIGATION OF GLOBAL CLIMATE CHANGE AND THE WORLD OF CHALLENGES

Arsenic groundwater contamination is a bane of human civilization and human scientific progress today. The world of challenges and the definite vision of environmental engineering techniques need to be revamped and re-organized with the passage of scientific vision, scientific rigor, and the timeframe. Global climate change and global warming are devastating the scientific landscape today. Bangladesh and India are on the verge of the world's largest environmental disaster such as lack of pure drinking water and severe health effects due to arsenic groundwater contamination. History of science and technology is today highly retrogressive as global environment stands at a dismal state of affairs. Technology management, engineering systems management, and human factor engineering are the ultimate answers to this monstrous environmental issue. The situation in West Bengal state of India and Bangladesh is grave and monstrous. Technology, engineering, and science have practically no answers to the ever-growing concerns of arsenic and heavy metal groundwater poisoning in India, Bangladesh, developing, and developed nations around the world. Environmental NT, technology management, and engineering systems management can solve vital problems of groundwater remediation and the vast domain of water purification. The participation of the civil society and the engineering domain is of utmost need for the emancipation of environmental and groundwater remediation. Nations around the world are highly concerned with climate change and global warming. The success and vision of chemical process engineering, environmental engineering, material science, and NT will surely open up the dawn of engineering and science in decades to come [12–14].

6.12 FUTURE RESEARCH TRENDS AND FUTURE FLOW OF THOUGHTS IN GROUNDWATER REMEDIATION

Groundwater remediation and drinking water treatment are the needs of humanity and scientific divinity in present-day human civilization. Future research trends should stress upon the application of NMs and engineered NMs in diverse areas of engineering and science, which definitely includes environmental engineering and chemical process engineering. The advancement of science in functionalized NMs in environmental remediation is the utmost need of the hour. The vision and the challenges of functionalized NMs applications are immense and far-reaching today. This paper is a veritable eye-opener towards the success of engineering and science globally today. Today human factor engineering and technology management should be integrated with every branch of science and engineering. The basic needs of human civilization and human scientific progress are immense, versatile, and vastly vital. Technology management and engineering systems management should also be linked to environmental protection and groundwater remediation. In this treatise, the author stresses with vast scientific rigor the targets of science and engineering, proper integrated water resource management and deep vision towards water resource engineering are the imminent needs of human society today. A big solution of groundwater remediation and arsenic groundwater decontamination today is integrated water resource management. Systems engineering approach and engineering systems management will go a long and effective way in the true realization of environmental engineering science and chemical process engineering today. Future research trends and future flow of thoughts in groundwater remediation should be targeted towards deep scientific inquiry and a larger vision of systems science and water resource management approach. It is then that the intricate questions of water purification, drinking water treatment, and industrial wastewater treatment are answered.

6.13 DETAILED RESEARCH RECOMMENDATIONS

Research recommendations in the field of environmental engineering and chemical process engineering are immense and ground-breaking today. Science and engineering of NT need to be re-envisioned with the passage of scientific history and time. Research frontiers in the field of environmental NT are vast, versatile, and far-reaching. NMs engineered NMs, and the vast world of environmental remediation are interconnected with each other in today's scientific scenario. Carbon nanotubes are highly used in manufacture and synthesis of nanoadsorbents in heavy metal removal from industrial wastewater. Buckminster fullerenes are the foundation stones of nanomaterial and engineered NMs research today. Technology and engineering science of NMs are slowly advancing and are today replete with vision and deep scientific rigor and inquiry. Research recommendations in the field of NT and environmental NT should be targeted towards greater emancipation in application areas in diverse areas of science and engineering [10, 11]. Chemical process engineering, energy engineering, and sustainability science are today in the path of newer scientific and engineering regeneration and scientific divinity [12–14]. The following are the detailed research recommendations:

- A greater scientific emancipation in the application of NMs in environmental remediation, the wider world of environmental engineering and energy engineering.

- A futuristic vision of health effects and health issues in the NMs and engineered NMs applications in human society.
- A vast scientific realization of groundwater remediation and engineered NMs applications.
- A deep scientific integration of adsorption and other novel separation processes such as membrane science.
- Removal of hazardous pollutants in industrial wastewater with the help of nanofiltration and other membrane separation processes.

The technology of nano-engineering, integrated water resource management, and human factor engineering needs to be re-envisioned and re-organized with the passage of scientific history and time. These branches of scientific endeavor need to be integrated if a greater emancipation of science needs to be realized. The other areas of sound research recommendations are membrane science, advanced oxidation processes, traditional, and non-traditional environmental engineering techniques. The questions of arsenic groundwater remediation techniques need to be addressed with immense effort and at war-footing in developing and developed nations around the world. Scientific contemplation and deep scientific consciousness should be the veritable futuristic pillars in the application of NT in environmental remediation. Thus the success of science and technology will usher in a newer era and a newer scientific regeneration.

6.14 FINDINGS ON THE STATE OF SCIENCE AND ENGINEERING

The strong findings of science and engineering in environmental protection need no barriers globally today. Water purification, drinking water hiatus, and industrial wastewater treatment stands in the midst of deep scientific provenance and vast scientific revival. Human civilization's vast scientific regeneration, mankind's deep scientific girth, and perseverance and the futuristic vision of environmental engineering will surely unravel the futuristic scientific inquiry and thoughts of science and engineering globally. The world of environmental engineering, chemical process engineering, and NT are facing immense scientific challenges. In many nations around the world, environmental sustainability stands in the midst of deep scientific provenance and vast ingenuity. Civil society participation in environmental engineering applications to humanity is the utmost need of the hour. The state of science and technology in the global scientific scenario is extremely dismal as global warming, and global climate change devastates the strong scientific fabric of might and vision. Chemical process engineering, environmental engineering science, and NT are the frontiers of science and engineering today. Provision of basic human needs such as energy, food, shelter, water, and education are at an immense state peril today. Here comes the importance of energy and environmental sustainability to humanity today. Science and engineering are two huge colossi with a definite vision of its own. Governments and civil society around the world should take tremendous scientific measures in mitigating global environmental issues, which includes the provision of clean water to human society. Technology and engineering science have practically no answers to the monstrous issues of groundwater heavy metal contamination. The state of science and engineering will surely be emboldened if civil society takes active participation in the true emancipation of energy and environmental sustainability.

6.15 CONCLUSION AND SCIENTIFIC PERSPECTIVES

Human civilization, human scientific progress, and global academic rigor in the field of NT and environmental protection are in the path of newer scientific rejuvenation. Developing and developed nations around the world are in the disastrous throes of the world's largest environmental disaster- arsenic groundwater and drinking water contamination. Thus human civilization is in the threshold of an unending scientific catastrophe. Technology, engineering, and science need to revamp their scientific perspectives as science and engineering surge forward towards a newer visionary era. Human factor engineering, systems science and industrial engineering need to be integrated with environmental protection and NT for the definite furtherance of global science and technology. The vast scientific challenges and the vision to excel are the forerunners towards a newer era in the field of water resources engineering, water resource management, sustainability, and environmental remediation. Environmental sustainability is the pillar of human futuristic vision in environmental protection today. A rigorous and concerted effort by scientists, engineers, and the civil society will surely pave the way towards larger emancipation of environmental protection today. In this entire chapter, the author pointedly focuses on the larger vision of environmental NT and the needs of humanity today. Environmental sustainability today stands in the midst of deep scientific and academic rigor. The proper implementation of environmental sustainability to human society is the utmost need today. In this entire treatise, the vision and the challenge of science and technology of NT and environmental protection are enumerated in deep details. The status of the environment is extremely dismal and thought-provoking today as science gears forward. Today environmental engineering stands on a solid foundation of integrated water resource management and systems engineering. The success, the targets, and the futuristic vision of water resource engineering, integrated water resource management, and environmental NT are enumerated with a vast vision of environmental sustainability. This treatise will surely be an eye-opener to the students and the teachers of NT as environmental engineering witness's immense scientific upheaval and deep scientific revival. The areas of nanocomposites applications will surely widen the scientific redeeming process. The challenge of science and engineering in the global scenario are dealt with immense insight and lucidity in this chapter. Technology will surely overcome the environmental engineering barriers of modern science and open up new avenues of research pursuit in years to come.

KEYWORDS

- environment remediation
- nanotechnology
- sustainability
- water resource engineering
- water resource management

REFERENCES

1. Arfin, T., & Tarannum, T., (2018). Engineered nanomaterials for industrial applications: An overview, Chapter-6. In: Chaudhery, M. H., (ed.), *Handbook of Nanomaterials for Industrial Applications* (pp. 127–134). Elsevier Inc., Amsterdam, Netherland.
2. Sen, R., & Chakrabarti, S., (2009). Biotechnology-applications to environmental remediation in recourse exploitation. *Current Science, 97*(6), 768–775.

3. Rajan, C. S., (2011). Nanotechnology in groundwater remediation. *International Journal of Environmental Science and Environment, 2*(3), 182–187.
4. Monteiro-Riviere, N. A., & Orsiere, T., (2007). Toxicological impacts of nanomaterials, Chapter 11. In: Wiesner, M. R., & Bottero, J. V., (eds.), *Environmental Nanotechnology-Applications and Impacts of Nanomaterials* (pp. 395–444). McGraw Hill Inc, New York, USA.
5. Mansoori, G. A., Bastami, T. R., Ahmadpour, A., & Eshaghi, Z., (2008). Chapter-2, environmental application of nanotechnology. In: Cao, G., & Brinker, C. J., (eds.), *Annual Review of Nano Research* (Vol. 2, pp. 439–489). World Scientific Publishing Company, Singapore.
6. *Report of the National Nanotechnology Initiative Workshop*, (2009). Arlington, Virginia, USA, National Science and Technology Council, USA.
7. Palit, S., & Hussain, C. M., (2018). Engineered nanomaterial for industrial use, Chapter 1. In: Chaudhery, M. H., (ed.), *Handbook of Nanomaterials for Industrial Applications* (pp. 3–12). Elsevier, Inc, Amsterdam, Netherlands.
8. Palit, S., (2018). Recent advances in the application of engineered nanomaterials in the environment industry: a critical overview and a vision for the future, chapter-47. In: Chaudhery, M. H., (ed.), *Handbook of Nanomaterials for Industrial Applications* (pp. 883–893). Elsevier Inc., Amsterdam, Netherlands.
9. Pathakoti, K., Manubolu, M., & Hwang, H. M., (2018). Nanotechnology applications for environmental industry, Chapter 48. In: Chaudhery, M. H., (ed.), *Handbook of Nanomaterials for Industrial Applications* (pp. 894–902). Elsevier Inc., Amsterdam, Netherlands.
10. Elimelech, M., & Phillip, W. A., (2011). The future of seawater desalination: Energy, technology and the environment. *Science, 333*, 712–717.
11. Palit, S., & Hussain, C. M., (2018). Frontiers of application of nanocomposites and the wide vision of membrane science: A critical overview and a vision for the future, Chapter 14. In: Hussain, C. M., & Mishra, A. K., (eds.), *Nanocomposites for Pollution Control* (pp. 441–476). Pan Stanford Publishing Pte. Ltd., Singapore.
12. *United States Environmental Protection Agency Report*, (2006). *In situ* treatment technologies for contaminated soil, Soild Waste and Emergency Response, (No.5203P).
13. Zhang, W. X., (2003). Nanoscale iron particles for environmental remediation: An overview. *Journal of Nanoparticle Research, 5*, 323–332.
14. Shannon, M. A., Bohn, P. W., Elimelech, M., Georgiadis, J. G., Marinas, B. J., & Mayes, A. M., (2008). *Science and Technology for Water Purification in the Coming Decades* (Vol. 452, pp. 301–310). Nature Publishing Group, London, United Kingdom.

IMPORTANT WEBLINKS

https://clu-in.org/techfocus/default.focus/.Environmental_Remediation/./Overview/ (Accessed on 18 July 2019).
https://en.wikipedia.org/wiki/Environmental_remediation (Accessed on 18 July 2019).
https://en.wikipedia.org/wiki/Nanotechnology (Accessed on 18 July 2019).
https://engineering.purdue.edu/EEE/Research/EnvironmentalRemediation (Accessed on 18 July 2019).

https://pdfs.semanticscholar.org/e420/20d98afedbea8d91b002cddea2dc6313a432.pdf
(Accessed on 18 July 2019).

https://sc-s.si/joomla/images/Environmental%20remediation.pdf (Accessed on 18 July 2019).

https://www.azonano.com/article.aspx?ArticleID=1134 (Accessed on 18 July 2019).

https://www.colorado.gov/pacific/cdphe/.and./environment/environmental-cleanup
(Accessed on 18 July 2019).

https://www.erfs.com/ (Accessed on 18 July 2019).

https://www.ghd.com/./what-s-shaping-the-future-of-environmental-remediation--.as
(Accessed on 18 July 2019).

https://www.iaea.org/sites/default/files/publications/magazines/./43205682024.pdf
(Accessed on 18 July 2019).

https://www.nano.gov/nanotech-101/what/definition (Accessed on 18 July 2019).

https://www.newcastle.edu.au › Centres › Environmental Remediation (GCER) (Accessed
on 18 July 2019).

https://www.researchgate.net › Engineering › Environmental Engineering (Accessed on 18
July 2019).

https://www.src.sk.ca/services/environmental-remediation (Accessed on 18 July 2019).

https://www.usaid.gov/vietnam/environmental-remediation-process (Accessed on 18 July
2019).

www.insituarsenic.org/details.html (Accessed on 18 July 2019).

www.mfe.govt.nz/more/environmental-remediation-projects (Accessed on 18 July 2019).

www.nnin.org/news-events/spotlights/what-nanotechnology (Accessed on 18 July 2019).

www.understandingnano.com/introduction.html (Accessed on 18 July 2019).

CHAPTER 7

Transport Coefficients in Multicomponent Systems: Mutual Diffusion, Self-Diffusion, and Tracer Diffusion

ANA C. F. RIBEIRO

Chemistry Center, Department of Chemistry, University of Coimbra, 3004-535 Coimbra, Portugal, E-mail: anacfrib@ci.uc.pt

ABSTRACT

Mass transport in electrolyte solutions of multicomponent systems is described having in mind the thermodynamics of irreversible processes. The concepts of mutual diffusion, self-diffusion, and tracer diffusion are discussed. In addition, the experimental methods in determination of these transport coefficients are presented and analyzed.

7.1 INTRODUCTION

The transport processes in the aqueous solutions result from a situation in which these systems are far from their thermodynamic equilibrium states, caused, for example, by heterogeneities of temperature, composition, or electrical potential. In each case, the flow of a system variable, J, is associated with the corresponding macroscopic quantity gradient, X.

$$J = -K \operatorname{grad} X \tag{1}$$

where K represents the transport coefficient (e.g., the mutual diffusion coefficient for mass transport [1]). Through the relation obtained by

Fourier [1] and the similarity established initially by Berthollet between diffusion and heat flow [1], Fick established that the force responsible for the flow in the process of diffusion in a binary system is a concentration gradient [1]. The measurements of transport properties require laborious attention and particular attention to possible sources of error, so that they can be obtained with the accuracy and precision required by science and technology. Among them, it is of interest the study of mutual diffusion coefficients [2]. We can say that the force responsible for diffusion is the chemical potential gradient of the diffusing substance, which is quantified in ideal solutions by the concentration gradient at a constant temperature [1]. Thus, two approaches are described to describe the isothermal diffusion of electrolytes in aqueous solutions: the laws of Fick and thermodynamics applied to irreversible processes. This last approximation allows to relate the diffusion coefficients, D, with other transport coefficients (denoted by intrinsic mobility, l_{ij}/N or l_{jj}/N [1–16]).

A generic treatment of mass transport processes in electrolyte solutions can be done by using the thermodynamics of irreversible or non-equilibrium phenomena. Miller [11–16] considered non-equilibrium thermodynamics a linear extension of classical thermodynamics. It is based on the hypothesis that classical concepts, such as temperature and entropy, are valid in small subvolumes which are assumed to be in "local" equilibrium. The absence of perturbations in the local equilibrium is assumed when temperature, composition, or temperature gradients occur, as well as the transport of heat, matter, or electricity through each sub-volume. These hypotheses have been experimentally tested in many physical and chemical processes, in particular for transport processes in electrolyte solutions. This is a general linear description of all types of transport processes. Specifically, it takes into account the effects of each type of flow on all others. In other words, the application of thermodynamics carried out in the works of Gibbs, and later Lewis and Randall deal almost entirely with states of equilibrium, and with reversible transitions between them [1]. Such transitions occur in a continuous set of equilibrium states by infinitely slow processes and can be reversed by an infinitely small variation in the variables that determine the nature of the system. Thus, classical thermodynamics can be successfully applied to some systems in which irreversible processes occur. More generally, if irreversible processes occur sufficiently slowly at the boundaries between homogeneous phases so that the mechanical and thermal equilibrium remains within the phases, the multiphase system can be treated thermodynamically. For homogeneous systems, the problem is solved by the local equilibrium hypothesis. The system is assumed to be divided into the identical number of subsystems, macroscopically small but microscopically large, each at constant volume and in internal thermodynamic equilibrium. If this procedure is valid, it is possible to calculate the rate of entropy production in homogeneous and heterogeneous systems (dS_{int}/dt) subject to irreversible processes. In other words, the rate of entropy production can be represented by a sum of products of the density of flux J_i and generalized and independent "thermodynamic forces" X_i. This product is referred to as the dissipation function, expressing, for an n-component system, by:

$$\emptyset = T\frac{dS_{int}}{dt} = \sum = J_i X_i \qquad (2)$$

where T is the temperature, J_i represents the fluxes of the ions or the solvent, and X_i is the thermodynamic force.

7.2 DIFFERENT CONCEPTS OF DIFFUSION

7.2.1 SELF-DIFFUSION

Many techniques are used to study diffusion in micellar solutions. It is very common to find misunderstandings concerning the meaning of a parameter, frequently just denoted by D and merely called diffusion coefficient, in the scientific literature, communications, meetings, or simple discussions among researchers. In fact, it is necessary to distinguish self-diffusion (intradiffusion, tracer diffusion, single-ion diffusion, ionic diffusion) and mutual diffusion (interdiffusion, concentration diffusion, salt diffusion) [1].

NMR and capillary-tube, the most popular methods, can only be used to measure intradiffusion coefficients [1.

NMR spectroscopy for solutions at thermodynamic equilibrium can be used together with the recently derived relations to predict multicomponent mutual diffusion D_{ik} coefficients for the nonequilibrium fluxes of solution component i driven by the gradient in concentration of component k.

7.2.2   MUTUAL DIFFUSION: BINARY, PSEUDO-BINARY, AND TERNARY SYSTEMS

The diffusion coefficient, D, in a binary system (i.e., with two independent components), may be defined in terms of the concentration gradient by phenomenological equations, known as Fick's first and second laws (Eqns. (3) and (4)) [1],

$$J_i = D_F \left(\frac{\partial c_i}{\partial x} \right) \tag{3}$$

$$\frac{\partial c}{\partial t} = \frac{\partial}{\partial x} \left(D_F \frac{\partial c}{\partial x} \right) \tag{4}$$

where J represents the flow of matter of component i across a suitable chosen reference plane per area unit and per time unit, in a one-dimensional system, and c is the concentration of solute in moles per volume unit at the point considered; D_F is the Fikian diffusion coefficient.

In reality, the gradient of chemical potential in the real solution must be taken as the true virtual force-producing diffusion and not the concentration gradient.

$$J_i = D_T \left(\frac{\partial \mu_i}{\partial X} \right) \tag{5}$$

where μ_i, μ_i^0 and f_i represent the potential and activity coefficient of component i, respectively, which definition of potential is given by Eqn. (6). Combining the Eqns. (5) and (6), and comparing the final result with the Eqn. (3), we obtain the relation between D_F and D_T, given by Eqn. (7).

$$\mu_i = \mu_i^0 + RT \ln f_i c_i \tag{6}$$

$$D_F = D_T \left(1 + \frac{\partial \ln f_i}{\partial \ln ci} \right) \tag{7}$$

This parameter, D_F, is not a pure kinetic parameter, because it depends on two contributions: kinetic (D_T) and thermodynamic ($\partial \mu / \partial c$, where μ represents the chemical potential) [1]. In other words, two different effects can control the diffusion process: ionic mobility and the gradient of the free energy.

However, as an approach, we can assume that the variation of the activity coefficient is not significant for the difference of concentration responsible by diffusion (Eqn. (7)), and, consequently, that for all practical purposes, D is a constant (and equal to the thermodynamic coefficient diffusion and the Fikian coefficient diffusion) (Eqn. (8)).

$$D_F = D_T \tag{8}$$

The open-ended capillary cell employed, which has previously been used to obtain mutual diffusion coefficients D_T for a wide variety of binary systems (e.g., electrolytes in aqueous solutions containing ions resulting from corrosion of dental material, such as chromium, and nickel), in different media [17–19]) has been described in great detail in previous papers [17–19]) (Figure 7.1). Basically, it consists of two vertical capillaries; each closed at one end by a platinum electrode, and positioned one above the other with the open ends separated by a distance of about 14 mm. The upper and lower tubes, initially filled with solutions of concentrations $0.75\ c$ and $1.25\ c$, respectively, are surrounded with a solution of concentration c. This ambient solution is contained in a glass tank ($200 \times 140 \times 60$) mm immersed in a thermostat bath at 25°C. Perspex sheets divide the tank internally, and a glass stirrer creates a slow lateral flow of ambient solution across the open ends of the capillaries. Experimental conditions are such that the concentration at each of the open ends is equal to the ambient solution value c, that is, the physical length of the capillary tube coincides with the diffusion path. This means that the required boundary conditions described in the literature [17] to solve Fick's second law of diffusion are applicable. Therefore, the so-called Δl effect [17] is reduced to negligible proportions. In our manually operated apparatus, diffusion is followed by measuring the ratio w = R_t/R_b of resistances R_t and R_b of the upper and lower tubes by an alternating current transformer bridge. In our automatic apparatus, w is measured by a Solartron digital voltmeter (DVM) 7061 with 6-1/2 digits. A power source (Bradley Electronic Model 232) supplies a 30 V sinusoidal signal at 4 kHz (stable to within 0.1 mV) to a potential divider that applies a 250 mV signal to the platinum electrodes in the top and bottom capillaries. By measuring the voltages V' and V" from the top and bottom electrodes to a central electrode at ground potential in a fraction of a second, the DVM calculates w.

In order to measure the differential diffusion coefficient D at a given concentration c, the bulk solution of concentration c is prepared by mixing 1 L of "top" solution with 1 L of "bottom" solution, accurately measured. The glass tank and the two capillaries are filled with c solution, immersed in the thermostat, and allowed to come to thermal equilibrium. The resistance ratio $w = w\infty$ measured under these conditions (with solutions in both capillaries at concentration c accurately gives the quantity $w\infty = 10^4/(1 + w\infty)$.

The capillaries are filled with the "top" and "bottom" solutions, which are then allowed to diffuse into the "bulk" solution. Resistance ratio readings are taken at various recorded times, beginning 1000 min after the start of the experiment, to determine the quantity $w = 10^4/(1 + w)$ as w approaches $w\infty$. The diffusion coefficient is evaluated using a linear least-squares procedure to fit the data and, finally, an iterative process is applied using 20 terms of the expansion series of Fick's second law for the present boundary conditions. The theory developed for the cell has been described previously [17].

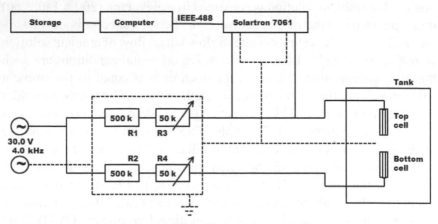

FIGURE 7.1   The open-ended capillary cell R1, R2, R3, R4: Resistances.

Concerning pseudo-binary systems, Ribeiro et al., [20–22], already studied systems such as $CuCl_2/\beta$-cyclodextrin/water [20], SDS/sucrose/water [21], $CuCl_2$/sucrose/water [22], using also the open-ended capillary cell. These systems are actually ternary systems, and we really have been measuring only main diffusion coefficients (D_{11}). However, from experimental conditions, we may consider these systems as pseudo-binary ones, and, consequently, take the measured parameters as binary diffusion coefficients, D.

Other experimental technique more useful because it permits to measure binary, ternary, and multicomponent diffusion is the Taylor dispersion method [23] (Figure 7.2). In a dispersion experiment, a slow, laminar flow of a liquid mixture is pumped through a long capillary tube, and a narrow pulse of a mixture of a slightly different composition is injected into this tube. The injected solutes spread out owing to the combined effects of convective flow and molecular diffusion. At the end of the diffusion tube, the dispersion is monitored by a flow-through detector (differential refractometer, ultraviolet-visible detector, etc.).

In the case of diffusion coefficients reported here, a Teflon tube of $3.2799 \pm (0.0001) \times 10^3$ cm-length, with a radius of 0.05570 (± 0.00003) cm, has been used. In a typical experiment, we have used the following procedure: at the start of each run, a 6-port Teflon injection valve (Rheodyne, model 5020) was used to introduce 0.063 cm^3 of solution into a laminar carrier stream of slightly different composition. A flow rate of 0.17 cm^3.min^{-1} was maintained by a metering pump (Gilson model Minipuls 3) to give retention times of about 8×10^3 s. The dispersion tube and the injection valve were kept at the work temperature ($\pm$ 0.01 K) in an air thermostat. Dispersion of the injected samples was monitored using a differential refractometer (Waters model 2410) at the outlet of the dispersion tube. Detector voltages, $V(t)$, were measured with a DVM (Agilent 34401 A) with an IEEE interface at accurately timed 5 s intervals. Binary diffusion coefficients were evaluated by fitting the dispersion equation (9) to the detector voltages.

$$V(t) = V_0 + V_1 t + V_{max} (t_R/t)^{1/2} \exp[-12D(t - t_R)^2/r^2 t] \qquad (9)$$

The additional fitting parameters were the mean sample retention time t_R, peak height V_{max}, baseline voltage V_0, and baseline slope V_1.

The concentrations of the injected solutions ($c_{av} + -c$) and the carrier solutions (c_{av}) differed by 0.004 mol dm^{-3} or less. Solutions of different composition were injected into each carrier solution to confirm that the measured diffusion coefficients were independent of the initial concentration difference and therefore represented the differential value of D at the carrier-stream composition.

Diffusion in a ternary solution is described by the diffusion Eqns. (10) and (11),

$$J_1 = -D_{11}\nabla C_1 - D_{12}\nabla C_2 \tag{10}$$

$$J_2 = -D_{21}\nabla C_1 - D_{22}\nabla C_2 \tag{11}$$

where J_1, J_2, $\partial c_1/\partial x$, and $\partial c_2/\partial x$ are the molar fluxes and the gradients in the concentrations of solute 1 and 2, respectively. Cross diffusion coefficients D_{12} and D_{21} give the coupled flux of each solute driven by a concentration gradient in the other solute. A positive D_{ik} cross-coefficient ($i \neq k$) indicates co-current coupled transport of solute i from regions of higher concentration of solute k to regions of lower concentration of solute k. However, a negative D_{ik} coefficient indicates counter-current coupled transport of solute i from regions of lower to higher concentration of solute k. Main diffusion coefficients give the flux of each solute produced by its own concentration gradient.

Extensions of the Taylor technique have been used to measure ternary mutual diffusion coefficients (D_{ik}) for multicomponent solutions. These D_{ik} coefficients, defined by Eqns. (10) and (11), were evaluated by fitting the ternary dispersion Eqn. (12) to two or more replicate pairs of peaks for each carrier-stream.

$$V(t) = V_0 + V_1 t + V_{max}(t_R/t)^{\frac{1}{2}} \left[W_1 \exp\left(-\frac{12 D_1 (t - t_R)^2}{r^2 t} \right) + (1 - W_1) \exp\left(-\frac{12 D_2 (t - t_R)^2}{r^2 t} \right) \right]$$

$$\tag{12}$$

Two pairs of refractive index profiles, D_1 and D_2, are the eigenvalues of the matrix of the ternary D_{ik} coefficients.

In these experiments, small volumes of ΔV of the solution, of composition $\overline{c_1} + \overline{\Delta c_1}$ and $\overline{c_2} + \overline{\Delta c_2}$ are injected into carrier solutions of composition, $\overline{c_1}$ and $\overline{c_2}$, at time $t = 0$.

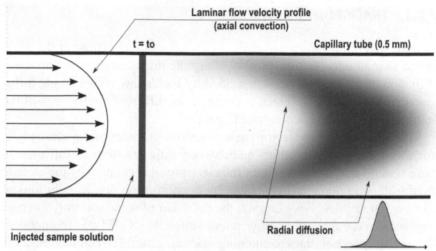

FIGURE 7.2 Schematic representation of the Taylor dispersion technique.

As examples, Ribeiro et al., have recently measured diffusion coefficients of some ternary systems, involving electrolytes and drugs in different media pH (e.g., ethambutol/HCl//water [24] and (Dopa/HCl/water [25]).

Succinctly, experimental methods that can be employed to determine mutual diffusion coefficients are diaphragm-cell (inaccuracy 0.5–1%), conductimetric (inaccuracy 0.2%), Gouy, and Rayleigh Interferometry (inaccuracy <0.1%) and Taylor dispersion (inaccuracy 1–2%). While the first and second methods consume days in experimental time, the last ones imply just hours. Concerning the conductimetric technique, despite this method has previously given us reasonably precise and accurate results, it is limited to studies of mutual diffusion in electrolyte solutions, and like diaphragm-cell experiments, the run times are inconveniently long (~days). The Gouy method also has high precision, but when applied to microemulsions, they are prone to gravitational instabilities and convection. Thus, the Taylor dispersion has become increasingly popular for measuring diffusion in solutions, because of its experimental short time and its major application to the different systems (electrolytes or nonelectrolytes). In addition, with this method, it is possible to measure multicomponent mutual diffusion coefficients.

7.2.3 *TRACER DIFFUSION*

In pure water, the anions and cations of an electrolyte diffuse at the same speed to maintain electroneutrality along the diffusion path. In this case, diffusion of the electrolyte is described by Fick's law with a single diffusion coefficient D, a weighted average of the diffusion coefficients of the ionic species and its counter-ions (Eqn. (3)).

The diffusion of an electrolyte in solutions of a supporting electrolyte, such as aqueous NaCl, differs qualitatively different from its diffusion in pure water. First, the diffusion of the electrolytes is coupled by the electric field (diffusion potential) generated by concentration gradients in ions of different mobilities. For example, the diffusion of some electrolytes (e.g., tetrasodium tetraphenyl porphyrin tetrasulfonate (Na4TPPS) (component 1)) in aqueous solutions containing sodium chloride (component 2) is described by the coupled Fick equations (10) and (11).

Cross-diffusion coefficients D_{12} and D_{21} are included for the flux of Na4TPPScaused by the sodium chloride concentration gradient (C_2) and the flux of sodium chloride caused by the Na4TPPSconcentration gradient (∇C_1).

In the limit $C_1/C_2 \rightarrow 0$ (a large molar excess of NaCl relative to Na4TPPS), D_{12} drops to zero because a NaCl (2) concentration gradient can't drive a coupled flow of Na4TPPS (1) in a Na_4TPPS-free solution. But cross-coefficient D_{21} is not zero, and is in fact quite large. This means the tracer diffusion of Na4TPPS in supporting NaCl solutions is described by the equations:

$$J_1(\text{Na4TPPS, tracer}) = -D_{11}\nabla C_1$$

$$J_2(\text{NaCl}) = -D_{21}\nabla C_1 - D_{22}\nabla C_2$$

If D_{21} is not zero, a concentration gradient in the Na4TPPS tracer will drive a significant coupled flow of NaCl in the carrier solution, producing ternary Taylor peaks (two overlapping Gaussian profiles).

However, from some experimental conditions and for some aqueous systems [26, 27] (e.g., amino acid (component 1) /NaCl (component 2), we are actually measuring the tracer diffusion coefficients D^0_{11} but not D^0_{12}, D^0_{21} and D^0_{22}. That is, the flow and injected solutions of compositions are $c_1 = 0$ and $c_2 = c_2$, and $c_1 = \Delta c$, $c_2 = c$, respectively, and the detector signal resembling a single normal distribution with variance $2t_R/24D_{11}$, and not two overlapping normal distributions. Thus, we may consider as an approach, the systems pseudo-binary and, consequently, take the measured parameters as the tracer diffusion coefficients of these amino acids in the aqueous solutions of this salt.

7.3 CONCLUSIONS

Mutual differential isotherm diffusion coefficients, D, self-diffusion, and tracer diffusion of electrolytes in aqueous solutions have been measured, having in mind a contribution to a better understanding of the structure of those solutions supplying the scientific and technological communities with data on these important parameters in solution transport processes. In fact, the scarcity of these parameters in the scientific literature, due to the difficulty on their accurate experimental measurement and impracticability of their determination by theoretical procedures, allied to the respective industrial and research need, well justifies efforts in such accurate measurements.

These data may be useful once they provide transport data necessary to model the diffusion for various chemical and pharmaceutical applications.

ACKNOWLEDGMENTS

The author is grateful for funding from "The Coimbra Chemistry Center" which is supported by the Foundation for Science and Technology (FCT), Portuguese Agency for Scientific Research, through the projects UID/QUI/UI0313/2013 and COMPETE.

KEYWORDS

- diffusion potential
- digital voltmeter
- mutual diffusion coefficients
- pseudo-binary systems
- refractometer
- thermodynamics
- tracer diffusion

REFERENCES

1. Tyrrell, H. J. V., & Harris, K. R., (1984). *Diffusion in Liquids: a Theoretical and Experimental Study.* Butterworths, London.
2. Lobo, V. M. M., (1990). *Handbook of Electrolyte Solutions.* Elsevier Sci. Publ., Amsterdam.
3. Miller, D. G., (1956). Thermodynamic theory of irreversible processes, I. The basic macroscopic concepts. *Am, J. Phys., 24,* 433.
4. Miller, D. G., (1960). Thermodynamic of irreversible processes. The experimental verification of the onsager reciprocal relations. *Chem. Rev., 60,* 15.
5. Miller, D. G., (1966). Application of irreversible thermodynamics to electrolyte solutions, I. determination of ionic transport coefficients l_{ij} for isothermal vector transport processes in binary electrolyte systems. *J. Phys. Chem., 70,* 2639.
6. Miller, D. G., (1967). Application of irreversible thermodynamics to electrolyte solutions, II. Ionic coefficients l_{ij} for isothermal vector transport processes in ternary electrolyte systems. *J. Phys. Chem., 71,* 616.
7. Miller, D. G., & Pikal, M. J., (1972). A test of the Onsager reciprocal relations and a discussion of the ionic isothermal vector transport coefficients l_{ij} for aqueous AgNO3 at 25°C. *J. Sol. Chem., 1,* 111.
8. Miller, D. G., (1978). Ionic interactions in transport processes as described by irreversible thermodynamics. Faraday discuss. *Chem. Soc., 64,* 295.
9. Dunsmore, H. S., Jalota, S. K., & Paterson, R., (1969). Irreversible thermodynamic parameters for isothermal vectorial transport processes in aqueous cesium chloride solutions. *J. Chem. Soc. (A), 7,* 1061.
10. Roessler, N., & Schneider, H., (1970). The transport properties of concentrated aqueous solutions of silver nitrate. *Ber. Buns. Phys. Chem., 74,* 1225.
11. Jalota, S. K., & Paterson, R., (1973). Irreversible thermodynamic parameters for isothermal transport processes in aqueous rubidium chloride. *J. Chem. Soc., Faraday Trans. I., 69,* 1510.
12. King, F., & Spiro, M., (1983). Transference numbers and phenomenological transport coefficients for concentrated aqueous hydrochloric acid solutions at 25°C. *J. Sol. Chem., 12,* 65.
13. Adhikari, M., Ghosh, D., & Chattopadhyay, P., (1985). A study on isothermal binary diffusion of some electrolytes in aqueous solution. *Ind. J. Chem. 24A,* 229.
14. Baabor, J., & Delgado, S. E. J., (1991). Irreversible thermodynamics of electrolyte solutions. Part III. Aqueous LiCl, NaCl and KCl solutions at 35°C. *Bol. Soc. Chil. Quím., 36,* 217.
15. Iadicicco, N., Paduano, L., & Vitagliano, V., (1996). Diffusion coefficients for the system potassium chromate-water at 25°C. *J. Chem. Eng. Data, 41,* 529.
16. Baabor, S. J., Gilchrist, M. A., & Delgado, E. J., (1995). Transport processes in the system LiCl-NaCl-H_2O at 30°C. An irreversible thermodynamic analysis. *Bol. Soc. Chil. Quím., 39,* 321.
17. Agar, J. N., & Lobo, V. M. M., (1975). Measurement of diffusion coefficients of electrolytes by a modified open-ended capillary method. *J. Chem. Soc., Faraday Trans., I71,* 1659.

18. Lobo, V. M. M., (1993). Mutual diffusion coefficients in aqueous electrolyte solutions. *Pure Appl. Chem., 65*, 2613.
19. Lobo, V. M. M., Ribeiro, A. C. F., & Verissimo, L. M. P., (1994). Diffusion coefficients in aqueous solutions of magnesium nitrate at 298 K. *Ber. Bunsenges Phys. Chem., 98*, 205.
20. Ribeiro, A. C. F., Esteso, M. A., Lobo, V. M. M., Valente, A. J. M., Simões, S. M. N., Sobral, A. J. F. N., et al., (2006). Interactions of copper (II) chloride with B-cyclodextrin aqueous solutions at 25°C. *J. Carbohydr. Chem., 25*, 173.
21. Ribeiro, A. C. F., Lobo, V. M. M., Azevedo, E. F. G., Da, M., Miguel, G., & Burrows, H. D., (2001). Diffusion coefficients of sodium dodecylsulfate in aqueous solutions and in aqueous solutions of sucrose. *J. Mol. Liq., 94*, 193.
22. Ribeiro, A. C. F., Esteso, M. A., Lobo, V. M. M., Valente, A. J. M., Simões, S. M. N., Sobral, A. J. F. N., & Burrows, H. D., (2007). Interactions of copper (II) chloride with sucrose, glucose and fructose in aqueous solutions. *J. Mol. Struct., 826*, 113.
23. Callendar, R., & Leaist, D. G., (2006). Diffusion coefficients for binary, ternary, and polydisperse solutions from peak-width analysis of Taylor dispersion profiles. *J. Sol. Chem., 35*, 353.
24. Verissimo, L. M. P., Ramos, M. L., Justino, L. L. G., Burrows, H. D., Cabral, A. M. T. D. P. V., Leaist, D. G., & Ribeiro, A. C. F., (2015). Ternary mutual diffusion in aqueous (ethambutol dihydrochloride plus hydrochloric acid) solutions. *J. Chem. Thermodyn., 90*, 140.
25. Barros, M. C. F., Ribeiro, A. C. F., Esteso, M. A., Lobo, V. M. M., & Leaist, D. G., (2014). Diffusion of levodopa in aqueous solutions of hydrochloric acid at 25°C. *J. Chem. Thermodyn., 72*, 44.
26. Leaist, D. G., (1991). Coupled tracer diffusion coefficients of solubilizates in ionic micelle solutions from liquid chromatography. *J. Solut. Chem., 20*, 175.
27. Rodriguez, D. M., Verissimo, L. M. P., Barros, M. C. F., Rodrigues, D. F. S. L., Rodrigo, M. M., Miguel, A. E., Romero, C. M., & Ribeiro, A. C. F., (2017). Limiting values of diffusion coefficients of glycine, alanine, α-aminobutyric acid, norvaline, and norleucine in a relevant physiological aqueous medium. *Eur. Phys, J. E., 40*(21), 1.

CHAPTER 8

Condensed Phosphates as Inorganic Polymers and Various Domains of Their Applications

M. AVALIANI,[1] V. CHAGELISHVILI,[1] N. BARNOVI,[1] N. ESAKIA,[2] M. GVELESIANI,[1] and SH. MAKHATADZE[1]

[1]Iv. Javakhishvili Tbilisi State University, Raphiel Agladze Institute of Inorganic Chemistry and Electrochemistry, 0186 Mindeli str., 11, Tbilisi, Georgia, E-mail: avaliani21@hotmail.com

[2]Iv. Javakhishvili Tbilisi State University, Faculty of Exact and Natural Sciences, Department of Chemistry, 0179 Chavchavadze ave. 3, Tbilisi, Georgia

ABSTRACT

In the 21st century, the chemistry of condensed phosphates, or that is to say inorganic polymers, developed quite intensively, for the cause of extensive use of phosphates materials in numerous fields of innovative technical domains. The importance of the efficient and resource-saving technologies for synthesis and applications of inorganic polymers/condensed compounds are not in doubt, which explains the relevance of our work, due to our technological methods and experience that we have developed and refined over many years. The presented data are the results of analysis, the examination of the experimental records, investigation, determination, and evaluation of properties of obtained compounds, notably of double condensed phosphates of polyvalent and monovalent metals. We have synthesized about 75 new condensed phosphates from solution-melts of polyphosphoric acids during the investigation of multi-component systems at the temperature range 400–850K. The investigation of condensed phosphates of alkaline and polyvalent metals for use as ion-exchange materials is importantly attractive and really interesting.

8.1   INTRODUCTION

Starting with the earliest work of the avant-gardist investigations of scientists, a large a number of condensed compounds—as a matter of fact, inorganic polymers—were synthesized around the world, and a great amount of advanced researche in 20th century was published and really esteemed [1–13]. In recent years, the various new biomaterials appeared on the base of condensed phosphates and/or polyphosphates. The investigation of condensed phosphates of alkaline earth metals for use as biomaterials is seriously attractive and encouraging [6]. It is worth noting the particular attention of scientists from different countries for the synthesis and study of the properties of double condensed phosphates of polyvalent metals. It is already known that these compounds, known as inorganic polymers [14–16], are widely used as thermo-resistant binders, glue compositions, and also as catalysts, ion-exchange materials [1, 4–6] as well as biomaterials [4–8, 10].

Very high thermal resistance and the thermo-stability, the vibrational and perfect luminescent characteristics of condensed phosphates determine their use in quantum electronics [1, 5, 9]. Multilateral spheres of application of condensed phosphates are very diverse: ion-exchange materials [1, 6], nanomaterials (NMs), efficient applying fertilizers, detergents, cement substances, catalytic agents, raw materials for phosphates glasses, thermo-resistant substances and also as food additive composites, in addition, the phosphates binding agents, phosphate-binders, and laser materials are supplanted/replaced by biomaterials, on the base of polyphosphates and hydroxylapatite [7–10]. On the strength of the varieties of phosphoric anions' condensation process, one of them leads to the predetermination of polymeric, cyclic or oligomeric structures of compounds, which finally permits the use of these compounds in various technical domains.

8.2   EXPERIMENTAL

This study is focused on the synthesis and analysis of some double condensed compounds of polyvalent and monovalent metals. In an glassy carbon crucible, there were mixed polyvalent metals oxides, ortho-phosphoric acid (percentage: 85%) and salts of monovalent metals in various molar ratio. Numerous condensed phosphates were prepared during the investigation of poly-component systems $M_2^I O - M_2^{III} O_3 - P_2 O_5 - H_2 O$ at the temperature range 400–850K. It is recognized that sufficient stability of polymeric phosphates in this respect makes it possible to identify and categorize them by the method of paper chromatography: to determine whether they are double di-, tri-, tetra-, octa-, and/or dodecaphosphates. This method, together with the chemical analysis, IR spectroscopy, thermogravimetric analysis, and X-ray diffraction (XRD) analysis, was used by us.

The presented data are the results of analysis, consideration of the experimental records, their examination, determination, and evaluation of properties of obtained compounds and correspondence with accomplishments and advancements in the area of inorganic polymer's chemistry [2–5, 9–10, 12, 13, 17–22], especially bearing in mind the fact that crystal structure was examined and described. The numerous compounds were wholly studied and observed by X-ray structural techniques. In addition were carried out investigations using an electronic scanning microscope of the Japanese company JEOL, equipped with a scanning electronic microscope JSM-6510LV (fitted by energy-disperse micro-roentgen spectral analyzer produced by Oxford Instruments'); The type of this analyzer is X-Max N 20. Electronic micrographs were carried out by means of reflected (BES) and as well as secondary (SEI) electrons at an accelerating voltage (at 20 kV). The working distance was approximately 15 mm, and micrographs have been taken at the diverse enlargements. The micro-spectroscopic analysis was performed from the sampling point zones and its surface area.

It should also be noted that we have studied the possibilities of application of mentioned phosphates [2, 5, 10, 15, 22–26].

8.3 RESULTS AND DISCUSSION

On the basis of many phosphates, especially condensed compounds of rare earth elements, luminophores, and materials for quantum electronics have been created. Many interesting works are devoted to the synthesis and investigation's studies of the mentioned condensed phosphates/inorganic polymers [1–4, 6–9, 11–13, 19–21].

Various diverse spheres of use of condensed phosphates are very different. In the frame of researches studies for novel compounds and new materials, certain condensed polyvalent metals phosphates having zeolite properties are well suited – due to its catalytic and adsorbent properties. Cyclic compounds with the general formula $M^IM^{III}P_nO_{3n}$ have free channels, and it can be predicted with certainty that they have zeolite properties. For example, in the cycles $Cs_3Ga_3P_{12}O_{36}$ and $Cs_3V_3P_{12}O_{36}$ the atoms of Ga and V atoms are built-in within the cycle [1, 2, 10, 13].

Figure 8.1 presents the structure of synthesized cyclododecaphosphate $Cs_3Ga_3P_{12}O_{36}$ [2, 4, 18].

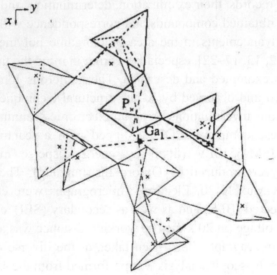

FIGURE 8.1　The structure of synthesized cyclododecaphosphate $Cs_3Ga_3P_{12}O_{36}$.

Due to the fact that the 12-member cycle is huge for them, (it is too large), they use only 6 oxygen atoms out of the total number (12), and the M^{3+} cations draw them to themselves, which actually leads to corrugation of the chain and reduces the free space of the channel.

Another is the situation in the case of $K_2M^{III}_2P_8O_{24}$ (where the trivalent metals are Al, Ga, Fe). It is undoubtedly warranted that cyclooctaphosphates $M^I_2M^{III}_2P_8O_{24}$ are the best catalysts in the process for receiving of olefin hydrocarbons and diene combinations. The assets of this compound in the model dehydration reaction of n-butyl alcohol by the impulsion method are investigated. It was founded on the conducted tests that the overall conversion was 52–65%. Studied sample is comparable to the activity of the sample BPO_3 (50%) obtained by mixing the starting components and surpassed the activity of the zeolite catalyst sample (32%), experienced under similar circumstances. It has also been found that the C_4 olefinic hydrocarbons were definitely recollected by the catalyst under the investigational situations.

The diameter of the channel in the case of gallium cyclooctaphosphate is 5.2 angstrom, which is an objectively good indicator. The $K_2Ga_2P_8O_{24}$ synthesized by us was studied for the catalytic activity (CA) at Moscow State University, the results showed that it has unique properties as an inorganic polymer and can be used as the best catalyst during organic synthesis reactions, specifically for the preparation of low molecular weight diene oléfines [1, 2, 10, 18].

The Infrared absorption spectra of some phosphates are presented below. IR absorption spectra frequency range is from 1700 to 300 cm^{-1} (see Table 8.1 and Figure 8.2).

TABLE 8.1 IR absorption Spectra Frequency for $K_2Ga_2P_8O_{24}$ and $Cs_3Ga_3P_{12}O_{36}$ (v, cm^{-1})

$K_2Ga_2P_8O_{24}$	$Cs_3Ga_3P_{12}O_{36}$
1310 strong	1312 strong
1225 strong	1250
1185 medium	1215 large strong
1125 medium	1164 medium
1050 large medium	1157 weak
782 weak	1112 weak
735 strong	1099 strong
635 strong	1084
595 strong	1064 strong
570 large strong	960 large strong
525 weak	880 medium
485 medium	790 medium
445 strong	730 strong
415 bending	700 medium
	614 strong
	590 strong
	557 weak
	497 strong
	470 large strong
	417 medium

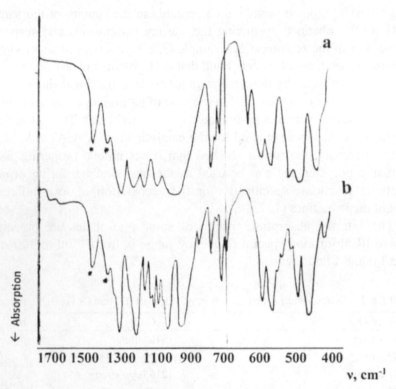

FIGURE 8.2   Infrared absorption spectra of $K_2Ga_2P_8O_{24}$ (a) and $Cs_3Ga_3P_{12}O_{36}$ (b).

As can be seen from the data given for $K_2Ga_2P_8O_{24}$ between frequency range 950–1050 cm^{-1} in absorption area v POP, a very large strong band appears. The absorption bands (1310 and 1225 cm^{-1}) are then quantitatively classified as of *strong*. It is not possible to determine the configuration of the anion.

In the IR spectrum of cesium-gallium cyclooctaphosphate (Figure 8.2), despite the high symmetry of the crystal-cubic symmetry – the strong splitting of the bands are observed in the area of phosphorus-oxygen valence bond vibrations (especially in the range 1040–1170 cm^{-1}). This is apparently associated with the presence of a great number of PO4 tetrahedrons in cyclic –dodeca-anion.

Many new interesting works about the application of condensed phosphates are published [6, 7, 10, 20, 25]. Inorganic phosphates are used in engineering and the fields of constructions to such an extent that it would be quite justified to mention on the industry supplying the necessary phosphate materials. The basics component of these materials is the so-called phosphate adhesive used in the form of acid phosphates of various metals, phosphoric acid or condensed forms, all mixed with polyvalent metal oxides forming autonomous masses. At present, there are many metallophosphate acidic adhesives, but on the outside among them, aluminum dihydrophosphate is widely used, aluminophosphate is more stable under storage conditions. The distinctive property of the hardened phosphate masses is that after an appropriate thermal treatment (between 100 and 500°C depending on the kind of the adhesive) becomes very stable and moisture resistant [9, 12]. The number of possible composition depends on the variation of conditions, for example, heating temperature, degree of dispersion and nature of filler, the acidity of the solution/melts, the molar ratio between the quantities of components to be mixed, etc. [12, 13, 17, 18].

Phosphates are also used in the frame of fine arts. This new trend in the art known as "thermophosphate pictorial art" is being developed in the Russian Federation by 0. Pavlov [19]. This author has developed mineral as well as phosphate paints in three forms: thermophosphate paints, powder colors, pastel, and artistic colors. The palette consists of 21 basic "rich" tints, and a pair of grades of each of them. The processes of the thermophosphate painting are very simple. The base material (asbestos—cement sheetings, fiber, cardboard, glass, ceramics, metal) is first coated with phosphate which is fixed by heating, then with phosphate colors (by a brush or by spraying). Thereafter, the color layer is treated/pulverized with a fixative containing phosphate adhesive, after it is heated during 2–3 minutes at 200–400°C [12, 10].

New interesting study about the application of polyphosphate with general formula $M(NH_4)_2(PO_3)_4$ as a fire retardant for polymeric material was published as Patent. The polymeric material is polyamide, glass-nylon composite, polycarbonate, or epoxy resin [20].

8.4 CONCLUSIONS

Numerous more of 75 condensed phosphates were prepared during the investigation of poly-component systems $M_2^I O - M_2^{III} O_3 - P_2 O_5 - H_2 O$ at the temperature range 400–850K.

During researches is established that synthesized by us the acidic triphosphates of Gallium and/or Indium $M^{III}H_2P_3O_{10} \cdot (1-2) H_2O$ is the finest ion-exchange agents, and this property proves the effectiveness of the application of mentioned compounds as ion exchangers. Analogically the similar acid triphosphate of Scandium $M^{III}H_2P_3O_{10} \cdot (1-2) H_2O$ maybe use as ion-exchange agents.

The comparison of the synthesized condensed compounds with similar and non-similar double phosphates, obtained by us earlier (in the systems containing many trivalent metals we can to summarize that while the radius of trivalent metal decreases, the polyphosphate chain identity period increases. The occurrence of this case is due to the complication of its form-factors.

Decrease of the correlation molar ratio enlarged the possibility of formation of great cycles, for example, for obtaining of cycloocta- or cyclododecaphosphates. The optimum achievement for the production of the great cyclic anions is parity of the big cations of monovalent metal against trivalent metals with small ionic radius.

ACKNOWLEDGMENTS

Finally, we would like to express our gratitude to our colleagues: Prof. Omar Mukbaniani, Dr. Dali Dzanashvili, and Eteri Shoshiashvili for their advice and support in presented work.

KEYWORDS

- **condensed phosphate**
- **cyclophosphate**
- **inorganic polymers**
- **ion-exchange materials**
- **phosphate glasses**
- **polyphosphate**

REFERENCES

1. Avaliani, M. A., Tananaev, I. V. N. V., & Gvelesiani, M. K., (2010). *Gaprindashvili. Chem. In-form Abstract: Issue Online. Synthesis and Investigation of Double Condensed Phosphates of Scandium and Alkali Metals.* DOI: 10.1002/ chin.199102035, Copyright © WILEY-VCH Verlag GmbH & Co. KGaA, Weinheim, online library.wiley.com/doi/10.1002.

2. Avaliani, M. A., (1982). *Condensed Phosphates of Gallium and Indium, PhD Thesis.* N. Kurnakov Institute of General and Inorganic Chemistry, Moscow, Russia.

3. Grunze, I., Palkina, K. K., Chudinova, N. N., Avaliani, M. A., & Guzeeva, L. S., (2009). Structure and thermal transitions of double Cs-Ca phosphates. OSTI ID: 5847982, *J. Inorg. Mater., 23*(4), 539–544.

4. Avaliani, M. A., & Tananaev, I. V., (2003–2011). Synthesis and investigation of double condensed phosphates. *Sc. J. Phosphorus, Sulfur Silicon Relat. Elem.* © FIZ CHEMIE Berlin.

5. Avaliani, M., (2004). Concerning applications of condensed phosphates of tri- and polyvalent metals. *Proceeding of the Georgian Academy of Sciences, 30*(1/2), 36–39.

6. Marsh, T. P., (2011). *Studies into the Ion Exchange and Intercalation Properties of AlH2P3O10·2H2O, PhD Thesis.* Etheses.bham.ac.uk/1599/1/University of Birmingham.

7. Hore, K., (2011). *The Investigation of Condensed Phosphates of Alkaline Earth Metals for Use as Biomaterials.* Res. Thesis. University of Birmingham.

8. Kulaev, I. S., Vagabov, V. M., & Kuakovskaya, T. V., (2005). *The Biochemistry of Inorganic Polyphosphates* (2nd edn.). John Wiley & Sons, Ltd.

9. Tananaev, I. V., (1980). *Pure & Appl. Chem. Review. Pergamon. Press Ltd., 52,* 1099–1115.

10. Avaliani, M., Purtseladze, B., Shohiashvili, E., Gvelesiani, M., & Barnovi, N., (2015). Apropos of inorganic polymers-condensed phosphates and spheres of their applications. *Proceeding of the Georgian Academy of Sciences, 41*(3), 227–231.

11. Abramovich, E. A., & Selevich, A. F., (2016). Thermal behavior of the system H_3BO_3–NH_4PO_3. *Sviridov Readings, Minsk, BSU. 12,* 26–31.

12. Tananaev, I. V., (1987). Some aspects of the chemistry of phosphates and their practical application. *Problems of Chemistry and Their Technology, Naüka.* Moscow, Russia.

13. Durif, A., (2013). *Crystal Chemistry of Condensed Phosphates* (p. 445). Springer Sci. & Business Media.

14. Mukbaniani, O. V., & Zaikov, G. E., (2003). *New Concepts in Polymer Science.* "Cyclolinear organosilicon copolymers: Synthesis, properties, application." Printed in Netherlands, ///VSP///, Utrecht, Boston.

15. Khananashvili, L. M., Mukbaniani, O. V., & Zaikov, G. E., (2006). *New Concepts in Polymer Science, "Elementorganic Monomers: Technology, Properties, Applications."* Printed in Netherlands, ///VSP///, Utrecht.

16. Khananashvili, L. M., & Mukbaniani, O. V., (1984). Macromolec. *Chem. Suppl., 6,* 77.

17. Palkina, K. K., Maksimova, S. I., Kuznetsov, V. G., & Chudinova, N. N., (1979). Structure of crystals of double octametaphosphate $Ga_2K_2P_8O_{24}$. *Doklady Akademii Nauk SSSR, [Crystallogr.], 245*(6), 1386–1389.

18. Avaliani, M., & Gvelesiani, M., (2006). Areas of crystallization of condensed scandium and cesium phosphates and regularities of their formation. *Proceeding of the Georgian Academy of Sciences, 32*(1/2), 52–58.

19. Pavlov, B., (1979). *Trans. of Ac. of Sci. of USSR, Series "Inorganic Materials," 15*(6), 985–988.

20. Balabanovich, A. I., Lesnikovich, A. I., & Selevich, A. F., (2017). *Flame Retardant for Polymeric Material*. Patent 20800.

21. Grunze, I., Palkina, K. K., Chudinova, N. N., & Avaliani, M. A., (1987). Structure and thermal rearrangements of binary cesium-gallium phosphates. *Chem. Inform, 18*(28), Wiley Online Library.

22. Avaliani, M., (**2015**). *General Overview of Synthesis and Properties of a New Group of Inorganic Polymers – Double Condensed Phosphates* (pp. 240–245). International conference on advanced materials and technologies (ICAMT- Tbilisi, Georgia).

23. Tananaev, I. V., & Avaliani, M. A., (1977). Synthesis of double gallium-potassium polyphosphates in polyphosphoric acid melts. Spring St, New York. *J. Inorganic Mater., 13*(12) 233

24. Avaliani, M. A., & Shapakidze, E., (2018). Areas of crystallization of double condensed phosphates of Ag and trivalent metals and regularities of their formation. 5th International conference on organic and inorganic chemistry (E-Poster). *Strategic Approach and Future Generation Advancements in Organic and Inorganic Chemistry.* Paris, France.

25. Avaliani, M., (2004). Special relevancy of practical use of condensed phosphates. *The 5th Republican Conference on Chemistry* (Vol. 1, pp. 15–16). Tbilisi, Georgia.

26. Avaliani, M., Shapakidze, E., Barnovi, N., Gvelesiani, M., & Dzanashvili, D., (2017). About new inorganic polymers-double condensed phosphates of silver and trivalent metals. *Journal of Chemistry and Chemical Engineering 11,* 60–64. doi: 10.17265/1934–7375/2017.02.004.

CHAPTER 9

Quantum Molecular Spintronics, Nanoscience, and Graphenes

FRANCISCO TORRENS[1] and GLORIA CASTELLANO[2]

[1]*Institute for Molecular Science, University of Valencia, PO Box 22085, E-46071, Valencia, Spain, E-mail: torrens@uv.es*

[2]*Department of Experimental Sciences and Mathematics, Faculty of Veterinary and Experimental Sciences, Valencia Catholic University Saint Vincent Martyr, Guillem de Castro-94, E-46001 Valencia, Spain*

ABSTRACT

Asimov foresaw *robots* but not miniaturization. Feynman predicted *miniaturization* and wrote a seminal report on it. There is plenty of room at the bottom, which results in an invitation to enter a new field of physics: the *nanoworld*. Did the invention of the *radio* revolutionize the world? No, it was not it but the *transistor radio receiver* that did it. This permitted that a shepherd in the top of a mountain could be in touch with what is going on. It would be interesting in surgery if one could swallow the surgeon like in *Fantastic Voyage*. A *space elevator* is a proposed type of planet-to-space transportation system. Size matters in the nanoscale. The materials of the 20th century were plastics. The materials of the 21st century are graphenes (GRs). Applications of GR in chemistry for a better life are energy, environment, health, and materials. The paradigm is now shifting as pure science opens new technology routes. The nanometer is the limit of chemical systems. Nanotechnology (NT) is a revolution in motion: electronics, image, materials, biomedicine, and energy. At the nanoscale, processes, e.g., energy production, catalysis, etc., can be optimized.

9.1   INTRODUCTION

On the one hand, Isaac Asimov (1950), in his science-fiction novels, predicted the future coming of robots but not the outcome of miniaturization: His robots were giant. On the other hand, Richard Feynman (1959) envisaged miniaturization and wrote a seminal report on it: "There's plenty of room at the bottom," which results an invitation to enter a new field of physics: the nanoscale world. Many broadcast professors say that the invention of the radio receiver revolutionized society. Notwithstanding, it was not it but its miniaturization, the transfer variator (transistor) radio, which transformed the world. For example, this invention allowed a shepherd in the top of a mountain to be connected with what is occurring now. As in Isaac Asimov (1966) novel "Fantastic Voyage," it would be interesting in surgery if one could miniaturize a surgeon in order to enter a patient's body. In another field of research, a space elevator was proposed as a planet-to-space transportation system.

In the nanoscale world, size is of importance in the nanometer range. The materials of the 20th century were polymer plastics. The materials of the 21st century are carbon nanographenes (GRs). Applications of GR in chemistry for a better life include energy, environment, health, and materials. The paradigm is now moving as pure science paves the way to new technology avenues. At the nanoscale, the nanometer is the limit of chemical systems because atoms constitute chemical entities. Nanotechnology (NT) is a revolution in motion: electronics, image, materials, biomedicine, and energy. At the nanoscale, processes, e.g., energy production, catalysis, etc., can be maximized, and full performance can only be achieved at this scale.

In earlier publications, it was reported that the periodic table of the elements (PTE) [1–3], quantum simulators [4–12], science, ethics of developing sustainability *via* nanosystems, devices [13], *green nanotechnology* as an approach towards environment safety [14], molecular devices, machines as hybrid organic-inorganic structures [15], PTE, quantum biting its tail and sustainable chemistry [16]. In the present report, it is discussed quantum molecular *spintronics* based on single-molecule magnets (SMMs), memory, spin quantum bits (*qubits*), Rabi nutation, nanoscience (NS), NT, the revolution of nanomaterials (NMs) and GRs. The aim of this work is to initiate a debate by suggesting a number of questions (Q), which can arise when addressing subjects of quantum molecular *spintronics*, SMM, NS, NT, the revolution of NMs and GRs, in different fields, and providing, when possible, answers (A) and hypotheses (H).

9.2 SMMS QUANTUM MOLECULAR SPINTRONICS: MEMORY/ SPIN QUBITS/RABI NUTATION

Spin electronics (*spintronics*), based on the freedoms of electron (e^-) charge, spin, and orbital, is a technology in the 21st century. Magnetic random access memory (MRAM), which uses giant (GMR) or tunneling (TMR) magnetoresistance, presents several advantages over conventional systems (e.g., nonvolatile information storage, high operation speeds on the order of nanoseconds, high storage densities, and low power consumption). Although the bulk of classical magnets composed of transition metal ions are normally used, Yamashita group used SMMs to overcome *Moore's Limitation* [17]. The SMMs underwent slow magnetic relaxation because of the double-well potential, defined as $|D|S^2$, and quantum tunneling, making them excellent materials for *quantum computers* and high-density memory storage devices. He presented single-molecule memory, spin *qubit*, and Rabi nutation at room temperature (RT). He informed SMMs encapsulated into single-wall C-nanotube (CNT) (SWCNT) to realize the new spintronics. They usually used the double-decker phthalocyaninato Tb^{III} SMM (TbPc$_2$) as a single-molecule memory. On the Au(111) substrate, they sublimated TbPc$_2$. By TMR *via* STM tip with one Co atom, they put the spin up and down on TbPc$_2$, and read them. As for the quantum computer, they synthesized 0–3-dimensional (D) V tetrakis (4-carboxyphenyl) porphyrin (TCPP) compounds. In 3D compounds, they realized spin qubit and observed Rabi nutation even at RT because of the rigid lattice. As for the new quantum molecular spintronics, they encapsulated SMMs into SWCNT. The SMMs behaviors were improved by the encapsulation into SWCNT.

9.3 NANOSCIENCE (NS) AND NANOTECHNOLOGY (NT): THE REVOLUTION OF NEW MATERIALS

García-Gómez proposed hypotheses, questions, and answers on NT as materials revolution [18].

H1. (Feynman, 1959). *There's plenty of room at the bottom* [19].

    Q1. How did the development of the techniques of study of NMs make possible applications?

    Q2. Which could evolution be in the future?

H2. Size matters: the nanometer (1nm).

H3. (Feynman, 1959). *There's plenty of room at the bottom.*

H4. (Feynman, 1959). *An invitation to enter a new field of physics.*

    Q3. What is a nanometer (nm)?

    A3. It is the thousand-millionth of a meter: $1nm = 10^{-9}m$.

    Q4. The size of an atom is 0.15nm, may something be smaller?

    Q5. What does it happen at the nanometric size?

    A5. The physical limit of the size of the atoms.

H5. (Drexler, 1986). *Engines of Creation: The Coming Era of Nanotechnology* [20].

    Q6. (Drexler, 1986). In addition, could we create life?

H6. (Zuart, 2003). Film: *Agent Cody Banks.*

H7. (Feynman, 1959). *It would be interesting in surgery if you could swallow the surgeon.*

H8. (Asimov, 1966). *Fantastic Voyage* [21].

H9. Materials of the 20th (plastics) and 21st (GR) century.

H10. *Space elevator.*

H11. Applications of GR. Chemistry for a better life: energy; environment; health; materials.

He provided the following conclusions (Cs).

    C1. The nanometer is the limit of chemical systems.

    C2. NT is a revolution in motion: electronics, image, materials, biomedicine, and energy.

9.4 GRAPHENES (GRS): THE MATERIALS OF THE 21ST CENTURY

Questions were raised on a roadmap for GR and Van Der Waals (VDW) heterostructures.

Q1. Could GR become the next disruptive technology, replacing used materials/leading to new markets?

Q2. Is it versatile enough to revolutionize many aspects of our life simultaneously?

Q3. How does the atomic-scale Lego® of VDW heterostructures work?

Q4. Why do VDW heterostructures deserve attention?

Q5.   What if one mimics layered semiconductors *via* atomic-scale Lego®?

Q6.   What if a dielectric plane (Bi/Sr/Ca/Cu oxide, BSCCO, hexagonal BN, hBN) is added in between intercalated-graphite planes?

Q7.   How to make VDW heterostructures?

Corma and García organized Areces Foundation International Symposium (València, Spain, 2015) on GRs. Müllen raised Qs on bottom-up synthesis [22].

Q8.   How to obtain carbon materials and GR?

Q9.   Size and solubility?

Su proposed questions and answer on GR/diamond hybrid materials for catalysis [23].

Q10.   How can carbon be chemically active?

A10.   Carbon can be chemically active by defects, bending, edges, and doping.

Q11.   The sp^2/sp^3 interface: structure and node?

Charola proposed Qs/A on technology for GR-materials production and applications [24].

Q12.   Existing applications?

Q13.   Overcapacity for now?

Q14.   Quantum dots (QDs)... how can things change dramatically?

Q15.   Synthesis of GR oxide (GO) Hammers' method uses toxins (H_2SO_4/$KMnO_4$/solvents), what can one do?

A15.   There is an improved Hummers' method for eco-friendly synthesis of GO.

Guinea proposed questions and answers on GR potential in microelectronics [25].

Q16.   Why are there two-dimensional (2D) crystals?

A16.   They are destabilized by fluctuations but, in the case of GR, at a distance of 1km.

Q17.   Where is twist-controlled resonant tunneling resonance coming from?

A17.   It is because of bubble formation.

Serp proposed Qs/As on few-layer GR (FLG) synthesis on transition metal catalysis [26].

Q18.   How to control carbon diffusion and precipitation?

Q19.   How to design catalyst structures in order to control more efficiently the catalytic process?

Q20.   How to obtain complex (rich) surface chemistry?

A20.   Oxidize with HNO_3.

Q21. Catalyst anchoring: how to do it?

Q22. Control of nanoparticle (NP) anchoring?

Q23. Control of NP size?

Q24. Control of NP location?

Q25. Control of NP organization (control of interparticle distance)?

A25. Control of catalyst synthesis at the molecular level.

Q26. Is FLG the best carbon catalyst or catalyst support?

A26. Yes, if good reactant adsorption exists near (e.g., *p*-chloronitrobenzene hydrogenation, Pd-catalyzed electrochemical alcohols oxidation).

Q27. Are CNTs too long?

A27. Mesodiffusion works at 2–3nm. The CNTs can be larger but can be cut.

Nguyen raised Qs on nanocomposites (NCs) of polymers and C-based nanofillers [27].

Q28. Can one apply synthesis to affect molecular-level chemical interactions and nanophenomena?

Q29. How to enhance the barrier properties of polymers?

Q30. Increasing nanofiller content?

Q31. Why have composites low strength?

Q32. How thin must the membrane be?

García-Gómez proposed Q/A on *filmogenic* biopolymers as GRs/2D-NMs precursors [28].

Q33. How is the B/C/N distribution after pyrolysis?

A33. They form islands.

Q34. How is GR spectroscopy?

A34. The problem is that GR is black.

Q35. How does GR compare with Pt catalyst for hydrogenation?

A35. Pt hydrogenates everything and is faster but not selective.

Q36. How does GR compare with metal Pd catalyst?

A36. Rising Pd composition increases catalytic activity (CA), but 0% extrapolation gives CA 8%.

Q37. What is the *filmogenic* biopolymer?

A37. It is chitosan (from chitin from crustacean exoskeleton) and alginate (from algae) [29]?

Q38. How is the O distribution in the graphitic material from chitosan and alginate?

A38. There is a 7–8% of O in the graphitic material and, theoretically, is concentrated in the periphery.

Q39. How is exfoliation carried out?

A39. With ultrasonication in any solvent.

Q40. What is the yield of the exfoliation?

A40. 70% (atomically thin layers) is stable in suspension, and 30% (multilayers) precipitate.

Q41. How can the quality of the graphitic material be controlled?

A41. The more pyrolytic temperature, the more loss of O, N, etc., arriving to pure C.

A hypothesis follows on a roadmap for GR.

H1. (Novoselov, 2012, Personal communication.). The paradigm is now shifting as pure science opens new technology routes.

9.5 DISCUSSION

Asimov foresaw robots but not miniaturization [30]. Feynman predicted miniaturization and wrote a seminal report on it. There is plenty of room at the bottom, which results in an invitation to enter a new field of physics: the nanoworld.

Communication professionals say that the invention of the radio revolutionized the world. However, it was not it but its miniaturization, the transistor radio receiver, which did it. For instance, this permitted that a shepherd in the top of a mountain could be in touch with what is going on. It would be interesting in surgery if one could swallow the surgeon like in Fantastic Voyage. A space elevator is a proposed type of planet-to-space transportation system.

In the nanoworld, size matters in the nanometer scale. The materials of the 20th century were plastics. The materials of the 21st century are GRs. Applications of GR in chemistry for a better life are energy, environment, health, and materials. The paradigm is now shifting as pure science opens new technology routes.

9.6 FINAL REMARKS

From the present results, the following final remarks can be drawn.

1. The nanometer is the limit of chemical systems because atoms constitute them.

2. NT is a revolution in motion: electronics, image, materials, biomedicine, and energy.
3. At the nanometer scale, processes, e.g., energy production, catalysis, etc., can be optimized.
4. Asimov foresaw robots, but Feynman predicted miniaturization.
5. Miniaturization, e.g., the transistor radio receiver, revolutionized the world.

ACKNOWLEDGMENTS

The authors thank support from Generalitat Valenciana (Project No. PROMETEO/2016/094) and Universidad Catolica de Valencia San Vicente Martir (Project No. 2019-217-001).

KEYWORDS

- graphenes
- microelectronics
- nanocomposite
- nanomaterial
- nanoparticle
- nanophenomenon
- nanoscale
- nanoscience

- nanosystem
- nanotechnology
- nanoworld
- quantum computer
- Rabi nutation
- single-molecule magnet
- spin *qubit*

REFERENCES

1. Torrens, F., & Castellano, G., (2015). Reflections on the nature of the periodic table of the elements: Implications in chemical education. In: Seijas, J. A., Vázquez, T. M. P., & Lin, S. K., (eds.), *Synthetic Organic Chemistry* (Vol. 18, pp. 1–15). MDPI: Basel, Switzerland.
2. Torrens, F., & Castellano, G., (2018). Nanoscience: From a two-dimensional to a three-dimensional periodic table of the elements. In: Haghi, A. K., Thomas, S., Palit, S., & Main, P., (eds.), *Methodologies and Applications for Analytical and Physical Chemistry* (pp. 3–26). Apple Academic–CRC: Waretown, NJ.

3. Torrens, F., & Castellano, G. (2019). Periodic table. In: Putz, M. V., (ed.), *New Frontiers in Nanochemistry: Concepts, Theories, and Trends* (Vol. 1, pp. 403–425). Apple Academic–CRC: Waretown, NJ.

4. Torrens, F., & Castellano, G., (2015). Ideas in the history of nano/miniaturization and (quantum) simulators: Feynman, education and research reorientation in translational science. In: Seijas, J. A., Vázquez, T. M. P., & Lin, S. K., (eds.), *Synthetic Organic Chemistry* (Vol. 19, pp. 1–16). MDPI: Basel, Switzerland.

5. Torrens, F., & Castellano, G., (2015). Reflections on the cultural history of nanominiaturization and quantum simulators (computers). In: Laguarda, M. N., Masot, P. R., & Brun, S. E., (eds.), *Sensors and Molecular Recognition* (Vol. 9, pp. 1–7). Universidad Politécnica de Valencia: València, Spain.

6. Torrens, F., & Castellano, G., (2016). Nanominiaturization and quantum computing. In: Costero Nieto, A. M., Parra, Á. M., Gaviña, C. P., & Gil, G. S., (eds.), *Sensors and Molecular Recognition* (Vol. 10, pp. 31–35). Universitat de València: València, Spain.

7. Torrens, F., & Castellano, G., (2018). Nanominiaturization, classical/quantum computers/simulators, superconductivity, and the universe. In: Haghi, A. K., Thomas, S., Palit, S., & Main, P., (eds.), *Methodologies and Applications for Analytical and Physical Chemistry* (pp. 27–44). Apple Academic–CRC: Waretown, NJ.

8. Torrens, F., & Castellano, G., (2018). Superconductors, superconductivity, BCS theory and entangled photons for quantum computing. In: Haghi, A. K., Aguilar, C. N., Thomas, S., & Praveen, K. M., (eds.), *Physical Chemistry for Engineering and Applied Sciences: Theoretical and Methodological Implication* (pp. 379–387). Apple Academic–CRC: Waretown, NJ.

9. Torrens, F., & Castellano, G., (2018). EPR paradox, quantum decoherence, qubits, goals and opportunities in quantum simulation. In: Haghi, A. K., (ed.), *Theoretical Models and Experimental Approaches in Physical Chemistry: Research Methodology and Practical Methods* (Vol. 5, pp. 317–334). Apple Academic–CRC: Waretown, NJ.

10. Torrens, F., & Castellano, G. (2019). Nanomaterials, molecular ion magnets, ultrastrong and spin-orbit couplings in quantum materials. In: Vakhrushev, A. V., Haghi, R., De Julián-Ortiz, J. V., & Allahyari, E., (eds.), *Physical Chemistry for Chemists and Chemical Engineers: Multidisciplinary Research Perspectives*, (pp. 181–190). Apple Academic–CRC: Waretown, NJ.

11. Torrens, F., & Castellano, G. (2019). Nanodevices and organization of single-ion magnets and spin qubits. In: Balköse, D., Ribeiro, A. C. F., Haghi, A. K., Ameta, S. C., & Chakraborty, T., (eds.), *Chemical Science and Engineering Technology: Perspectives on Interdisciplinary Research,* (pp. 67–74). Apple Academic–CRC: Waretown, NJ.

12. Torrens, F., & Castellano, G. (2019). Superconductivity and quantum computing via magnetic molecules. In: Mukbaniani, O. V., Balk-se, D., Susanto, H., Haghi, A. K., (eds.), Composite Materials for Industry, Electronics, and the Environment: Research and Applications (pp. 201-209). Apple Academic–CRC: Waretown, NJ.

13. Torrens, F., & Castellano, G. (2019). Developing sustainability via nanosystems and devices: Science–ethics. In: Balköse, D., Ribeiro, A. C. F., Haghi, A. K., Ameta, S. C., & Chakraborty, T., (eds.), *Chemical Science and Engineering Technology: Perspectives on Interdisciplinary Research.* (pp. 75–84). Apple Academic–CRC: Waretown, NJ.

14. Torrens, F., & Castellano, G. Green nanotechnology: An approach towards environment safety. In: Vakhrushev, A. V., Ameta, S. C., Susanto, H., & Haghi, A. K., (eds.), *Advances in Nanotechnology and the Environmental Sciences: Applications, Innovations, and Visions for the Future.* Apple Academic–CRC: Waretown, NJ, in press.

15. Torrens, F., & Castellano, G. Molecular devices/machines: Hybrid organic-inorganic structures. In: Pourhashemi, A., Deka, S. C., & Haghi, A. K., (eds.), *Research Methods and Applications in Chemical and Biological Engineering.* Apple Academic–CRC: Waretown, NJ, in press.

16. Torrens, F., & Castellano, G. The periodic table, quantum biting its tail, and sustainable chemistry. In: Torrens, F., Haghi, A. K., & Chakraborty, T., (eds.), *Chemical Nanoscience and Nanotechnology: New Materials and Modern Techniques.* Apple Academic–CRC, Waretown, NJ, in press.

17. Yamashita, M. *Personal Communication.*

18. García-Gómez, H. *Personal Communication.*

19. Feynman, R. P., (1960). There's plenty of room at the bottom. *Caltech Eng. Sci., 23,* 22–36.

20. Drexler, E., (1990). *Engines of Creation: The Coming Era of Nanotechnology.* Anchor: New York, NY.

21. Asimov, I., (1988). *Fantastic Voyage.* Bantam: New York, NY.

22. Müllen, K., (2015). *Book of Abstracts, International Symposium Graphenes: The Materials of the XXIst Century.* València, Spain, October 21, Fundación Ramón Areces: València, Spain, O-1.

23. Su, D. S., (2015). *Book of Abstracts, International Symposium Graphenes: The Materials of the XXIst Century.* València, Spain, October 21, Fundación Ramón Areces: València, Spain, O-2.

24. Charola, Í., (2015). *Book of Abstracts, International Symposium Graphenes: The Materials of the XXIst Century.* València, Spain, October 21, Fundación Ramón Areces: València, Spain, O-3.

25. Guinea, F., (2015). *Book of Abstracts, International Symposium Graphenes: The Materials of the XXIst Century.* València, Spain, October 21, Fundación Ramón Areces: València, Spain, O-4.

26. Serp, P., (2015). *Book of Abstracts, International Symposium Graphenes: The Materials of the XXIst Century.* València, Spain, October 21, Fundación Ramón Areces: València, Spain, O-5.

27. Nguyen, S. B. T., (2015). *Book of Abstracts, International Symposium Graphenes: The Materials of the XXIst Century.* València, Spain, October 21, Fundación Ramón Areces: València, Spain, O-6.

28. García-Gómez, H., (2015). *Book of Abstracts, International Symposium Graphenes: The Materials of the XXIst Century.* València, Spain, October 21, Fundación Ramón Areces: València, Spain, O-7.

29. Ratner, B. D., Hoffman, A. S., Schoen, F. J., & Lemons, J. E., (2013). *Biomaterials Science: An Introduction to Materials in Medicine.* Academic Press: New York, NY.

30. Asimov, I., (1950). *I, Robot.* Gnome, New York, NY.

CHAPTER 10

Characterization Techniques for Polymers and Polymer Nanocomposites

SHRIKAANT KULKARNI

Department of Chemical Engineering, Vishwakarma Institute of Technology, Pune (M.S.), India, E-mail: shrikaant.kulkarni@vit.edu

ABSTRACT

Polymeric materials in various forms contribute substantially to materials in general. They exhibit a difference in behaviors and functionalities than conventional materials. However, to understand the behavioral changes of polymers and polymer nanocomposites (NCs) their property profile has to be understood which reflects upon their chemistry, functionality, morphology, particle size, etc. which further decides the application for which it can be used. However, to get a better insight into its characterization using various analytical tools is a must. Different analytical tools are used for the internal and external elucidation of structure, and to understand the surface chemistry and thereby the behavior of the synthesized polymers and polymer NCs. A number of characterization tools which are put into use include thermogravimetric analysis (TGA), differential scanning calorimetry (DSC), transmission electron microscopy (TEM), scanning electron microscopy (SEM), x-ray diffraction (XRD), nuclear magnetic resonance (NMR), Fourier transform infrared spectroscopy (FTIR), Raman spectroscopy (RS), x-ray photoelectron spectroscopy (XPS), dielectric relaxation spectroscopy (DRS), atomic force microscopy (AFM), electron spin resonance (ESR), continuous wave and pulsed ESR spectroscopy, etc. to unravel properties like mechanical performance, fire behavior, barrier performance, biodegradability, rheological properties, PVT characterization, tribological behavior, etc. At times hyphenated or coupled techniques too are required to be used to characterize the material both qualitatively and quantitatively accurately. For example, the intercalated and exfoliated NCs are characterized using wide-angle x-ray diffraction (WAXRD) in conjunction with TEM for the examination of both intercalation and filler exfoliation together.

10.1 INTRODUCTION

Polymeric materials in various forms including a relatively new family of nanoscale materials with at least one dimension of the filler phase smaller than 100 nm [1–9] contribute to a major chunk of materials at large. They differ in behaviors and functionalities from that of conventional materials. However, to understand the behavioral changes of polymers, their property profile has to be understood, which in turn is a reflection of their chemistry, functionality, morphology, particle size, etc. The properties possessed by any material depends upon the chemistry (Structure, nature, and kind of constituent particles, kind, and magnitude of forces of attraction among the particles) of the material in question which ultimately decides the utility or the purpose for which the given material can be used in preference. However, to get a better insight into any material characterization of it using various analytical tools based on what sort of information is to be ascertained. For example, polar polymers have better filler dispersion than non-polar ones owing to better interfacial adhesion between the surface polarities of filler (reinforcing material) and polymers (matrix). To get over these difficulties, either low molecular weight interfacial adhesives are added to the system or the filler surface is functionally modified by using different chemical or physical processes. The different techniques to modify the surface chemistry of the filler as well as to synthesize the polymer nanocomposites (NCs) need to be accompanied by robust characterization to gain insights into the factors instrumental in the nanocomposite microstructure and properties so as to be able to design them in tune with the demand of the application. Therefore, characterization of morphology, properties of the polymeric materials is a prerequisite to analyze various facets of polymers [1–3]. A few of them are listed as follows for polymer NCs:

    a.   Quality, orientation or alignment of the filler of dispersion in the polymer matrix in addition to the processing method used;

b. Effect of filler surface chemistry on filler dispersion behavior and resulting composite properties;

c. Adhesion of the modified filler with the polymeric chain molecules including chemical affinity between the two phases;

d. Effect of variations in the process parameters on the resulting morphology and properties;

e. Analysis of a host of properties to unravel the application potential of the NCs [4].

It is also, at many instances, necessary to employ more than one characterization technique in order to accurately characterize the nanocomposite material, e.g., over the years, it has become common to divide the NCs into intercalated and exfoliated types based on the reflection patterns observed in the detection range of characterization tool like wide-angle X-ray diffraction (WAXRD). However, this classification is arbitrary because the observation of a peak in the diffractogram depends on the periodicity, concentration, and orientation of the aluminosilicates, but does not exclude the presence of exfoliated part. Its absence also does not exclude the presence of small or randomly oriented intercalated particles and, therefore, does not indicate complete exfoliation as often postulated. The extent of filler intercalation or exfoliation could not be quantified. However, when the same NCs characterized by transmission electron microscopy (TEM), extensive filler exfoliation can be observed. The intercalated platelets also had varying thicknesses. Thus, to generate better insights into the nanocomposite microstructures, synergistic coupling, or hyphenation of different characterization techniques is useful [5].

A number of different nanocomposite characterization tools are available which include thermogravimetric analysis (TGA), differential scanning calorimetry (DSC), TEM, scanning electron microscopy (SEM), x-ray diffraction (XRD), nuclear magnetic resonance (NMR), Fourier transform infrared spectroscopy (FTIR), Raman spectroscopy (RS), x-ray photoelectron spectroscopy (XPS), dielectric relaxation spectroscopy (DRS), atomic force microscopy (AFM), electron spin resonance (ESR), continuous – wave and pulsed ESR spectroscopy, etc.

Apart from that, numerous characterization techniques to ascertain nanocomposite properties like mechanical performance, fire behavior, barrier performance, biodegradability, rheological properties, PVT characterization, tribological behavior, etc., are also used [6].

10.2 CHEMICAL ANALYSIS OF POLYMERS

Chemical analysis of polymers and polymer-based products is tedious due to huge number and kind of such materials and because of modification and compounding of it.

For qualitative analysis of polymers, it is very important for the samples to be highly pure without additives such as plasticizers, fillers, or stabilizers incorporated into it. One must extract or reprecipitate additives before their identification. The solvents and precipitating agents are substance-specific and should be chosen selectively [7, 8].

The quantitative composition of a polymer, demands following sequence of operations:

- Comminution of the polymer sample.
- Recovery of additives.
- Qualitative and quantitative examination of the additives.
- Qualitative and quantitative analysis of the separated polymer samples.

10.2.1 MECHANICAL COMMINUTION, SEPARATION, AND IDENTIFICATION OF POLYMERS

Mechanical comminution: It involves the use of cutting tools such as shears, knives, or razor blades and Drilling, milling, etc. such that a smaller particle diameter can be obtained. The reproducibility and uniformity of the comminution processes are of paramount importance when subjected to mechanical loads or low-temperature treatments.

Comminution is followed by the conditioning of samples over P_2O_5 at room temperature (RT) in desiccators, e.g., Plasticizers (1 to 2 g) can be separated by extraction with diethyl ether in a Soxhlet apparatus. Stabilizers based on pure organic or organometallic compounds may be separated partially. After distilling off the ether followed by drying the extract at 105°C to constant weight, the amount of ether soluble components is calculated from the difference in weight before and after the extraction of the extraction flask. Plasticizers namely esters of a few aliphatic and aromatic mono and dicarboxylic acids, aliphatic, and aromatic phosphorus acid esters, ethers, alcohols, ketones, amines, amides, and non-polar and chlorinated hydrocarbons [9].

Thin-layer chromatography (TLC) is preferred for their separation and qualitative detection. Usually, Kieselgur plates (0.25 mm thick, at 110°C for 30 min) in the saturated vapor are used preferably. Methylene chloride and mixtures with diisopropyl ether at temperatures between 40 to 60°C have so far been successfully used as the solvents. TLC was the preferred separation method characterized by its high-resolution efficiency, rapidity, and variety of detection possibilities. Usually, 0.5 mm thick silica-gel-G-plates are used, activated at 120°C for 30 min. in a supersaturated atmosphere. However, HPLC in isolation or HPLC-MS like hyphenated techniques is used preferably over TLC because of several advantages of it [10].

Although selectivity of TLC can be increased with the optimization of method parameters, gas chromatography (GC) must be preferentially used for complex plasticizer mixtures. The gas-chromatographic separation of plasticizers can be done directly or after derivatization to low boiling compounds by transesterification Fillers can be quantitatively separated from polymers by centrifugation followed by decantation of the solvent. Carbon black as an exception cannot be separated totally, even under the most severe centrifugation and other specialized methods. Normal methods of qualitative inorganic analysis can also be used for filler identification. Quantitative determination is done gravimetrically [11].

10.2.2 QUALITATIVE AND QUANTITATIVE ANALYSIS OF POLYMERS

Polymer identification starts with a sequence of preliminary tests. Polymers are difficult to identify because of factors like the presence of copolymers, macromolecular properties, etc. unlike low molecular weight compounds, which can be identified using physical constants like melting or boiling points, etc. Further, the efficiency of physical methods such as IR, NMR spectroscopy, as well as GC on pyrolysis, will have an added advantage.

Pyrolysis involves about heating 0.1 g of the sample carefully in a 60 mm long glow tube with a diameter of 6 mm over a small flame. Depolymerization is a special case of thermal degradation. This can be seen very clearly in PVC and PVAC. Depolymerization, elimination, and statistical chain-scission reactions can be used for polymer analysis. When the monomer is the main degradation product, it can easily be identified from physical properties like boiling point and refractive index. Elimination and chain-scission reactions generate characteristic pyrograms, which can be followed by GC or IR spectroscopy. For testing the depolymerization behavior of a typical polymer sample (0.2 to 0.3 g) is heated gently to a temperature of 500°C in a distillation flask. The boiling point and refractive index of the distillate are then determined [12, 13].

10.2.3　QUALITATIVE AND QUANTITATIVE ANALYSIS OF STABILIZERS

Stabilizers like heat, light, and oxidation-resistant have acquired a great importance. Different examples of stabilizers are shown in Table 10.1.

TABLE 10.1　Representative Examples of Stabilizers

Stabilizers	Heat absorbers	UV Light Absorbers	Antioxidants
Examples	Salts of heavy metals, metal salts of organic acids, nitrogenous organic compounds	Hydroxy benzophenone derivatives, salicyl esters, and benzotriazoles	Phenols, aromatic amines, and benzimidazoles

Qualitative analysis of stabilizers is cumbersome because of their huge number and amounts in traces, vulnerability to undergo transfer or rearrangement reactions. Their detection is of importance because of the hazardous nature of its decomposition products. Stabilizers are identified either by solid-liquid extraction or by precipitation from a dilute solution [14].

10.3　CHROMATOGRAPHIC TECHNIQUES

10.3.1　CHROMATOGRAPHY

1. **Principle:** The separation of molecular mixtures by the distribution of solutes between two or more phases, one phase being essentially stationary two-dimensional (a surface) and the other phase, being a mobile bulk phase brought into contact in a counter-current fashion with the stationary phase.
2. **Theory:** Chromatography using samples in different physical states like solid, liquid, gas, and plasma, supercritical fluid are available.

The sequences of steps in chromatographic separation are as follows:

- A sample is placed at the top of a column where its components are sorbed and desorbed by a carrier during its downward journey.
- This partitioning process takes place repeatedly as the sample traverses towards the end of the column.
- Each solute migrates at its own rate through the column, consequently, to form a band (peak) representing the solute on the column.
- A detector responds to each band representing the solute. The output of detector response versus time is called a chromatogram.
- The retention or emergence time is the characteristic of the component.
- The peak area is a measure of concentration, based on the calibration curve with the pure compound as a reference [15].

10.3.2 GAS CHROMATOGRAPHY (GC)

1. **Principle:** Thermally stable and vaporizable or volatile samples are separated using gas as the moving phase, and a solid or a nonvolatile liquid as a stationary phase in the technique of GC.
2. **Theory:** In GC, the sample is usually injected at a programmed temperature to ensure vaporization. Obviously, only samples which volatilize at the given temperature can be analyzed [16] (Figure 10.1).

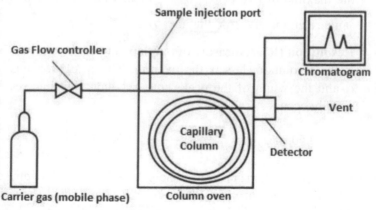

FIGURE 10.1   Gas chromatograph.

10.3.2.1 GC PARAMETERS

- **Retention Time (t_R):** The time an analyte takes to move through the column or the amount of time an analyte spends in the column or sum of the time spent in the stationary phase and the mobile phase.

- **Retention Time of Unretained Compound (t_M):** The time an unretained compound takes to run through the column or the rate at which the mobile phase (carrier) journeys down the column, or it is equivalent to the time an analyte spends in the mobile phase.

- **Retention Factor (k):** Ratio of the amount of time an analyte spends in the stationary and mobile phases, i.e., $K = \dfrac{t_R - t_M}{t_M}$

- **Distribution Coefficient (K):** Ratio of analyte concentration in the stationary phase and mobile phase, i.e., $K = \dfrac{C_S}{C_M}$ where c_S, $c_M =$ concentration of an analyte in the stationary and mobile phase, respectively.

- **Phase Ratio (β):** The phase ratio refers to a compound's retention, for the given stationary phase and at a given column temperature (program or isothermal). An increase in phase ratio results in a decrease in retention (k), since K is constant; and vice versa, i.e., $K = k\beta$; $\beta = r/2d_f$ where r = column radius (μm) and d_f = film thickness (μm).

- **Separation Factor (α):** A measure of the time or distance between the maxima of two peaks, $\alpha = 1$ means the two peaks have the same retention and co-elute, i.e., $\alpha = \dfrac{k_2}{k_1}$

- **Resolution (R):** A measure of overlap between two peaks; higher is the resolution, less is the overlap. The resolution takes both α and the width of the peaks into account. Baseline resolution usually occurs at:

$$R = 1.50, \text{ i.e., } R = 2\left[\frac{t_{R2} - t_{R1}}{W_{b1} + W_{b2}}\right] \text{ or } R = 1.18\left[\frac{t_{R2} - t_{R1}}{W_{h1} + W_{h2}}\right]$$

where w_h = peak width at half peak height, and w_b = peak width at base

- **Number of Theoretical Plates (N):** It is an indirect measure of peak width for a peak at a specific retention time. Columns with high N have a narrower peak are considered to be more efficient than those with lower N. Column efficiency is a function of Column dimensions, type of carrier gas and its average linear velocity, compound, and its retention, i.e., $N = 5.545 \dfrac{t_R}{W_h}$ or $16 \dfrac{t_R}{W_h}$.

- **Height Equivalent to a Theoretical Plate (HETP):** Another measure of column efficiency. Small plate heights indicate higher efficiency, i.e., $H = \dfrac{L}{N}$, where L = column length (mm) and N = theoretical plates number.
- **Carrier Gas Linear Velocity (v):** Affects the chromatographic resolution (i.e., separation efficiency). Each gas has a linear velocity at which optimum separation can be achieved (smallest HETP).

10.3.2.2 TYPICAL APPLICATIONS OF GC

Following are some of the representative applications of GC:

- Pesticide residues and pollutants in water, agricultural products, foodstuff.
- Organic solvents in packaging materials, ink, etc.
- Drugs of abuse in urine, blood, tablets.
- Fatty acid contents in edible oils, fat, etc.
- Essential oil.

10.3.2.3 TYPES OF GC

GC has developed into one of the most powerful analytical tools available to the organic chemist [17].

1. **Gas-Solid Chromatography:** If the stationary phase is solid, the technique is called as gas-solid chromatography. Adsorption is the principle underlying separation. The analytes which are held strongly stay longer in the stationary phase and are separated later.
2. **Gas-Liquid Chromatography:** If the stationary phase is a liquid, the technique is gas-liquid chromatography, and the partition or solubility difference between the solutes in the liquid phase is the principle of separation. GC provides for separation of minute quantities of analytes, flexibility in terms of separation broad

boiling ranges using either packed filled or capillary columns. The detector used to sense and quantify the specificity and sensitivity also offers versatility in sample analysis.

10.3.3 *LIQUID CHROMATOGRAPHY (LC)*

1. **Principle:** Thermally labile and volatile/nonvolatile samples are separated using the liquid as the moving phase, and a solid as a stationary phase in the technique of liquid chromatography (LC).
2. **Theory:** In LC, the sample is first dissolved in the moving phase (solvent) and injected at ambient temperature. Volatility of the sample is not a prerequisite. The right choice of solvent for the right sample is what is required for the complete dissolution of the sample. Therefore, LC is advantageous over GC:

The advancement in LC has taken place from classical open column to the instrument-based high-performance liquid chromatography (HPLC) and Ultra performance or Ultrafast LC for analytical as well as preparative work. It is characterized by high sensitivity, resolution, speed, sensitivity, and low dead volume detectors.

10.3.3.1 *TYPES OF LC*

There are four different types of LC, based on the nature of the stationary phase and the separation mechanism.

*   Liquid/liquid chromatography (LLC) is a partition chromatography. The sample is retained by partitioning between mobile liquid and stationary liquid. e.g., paper chromatography.
*   Liquid/solid chromatography (LSC) is an adsorption chromatography. Adsorbents such as alumina and silica gel are packed in a column, and the samples components are migrated by a mobile phase, e.g., TLC and column chromatography.
*   Ion-exchange chromatography (IEC) employs ion exchangers such as zeolites and synthetic organic and inorganic resins to bring about the chromatographic separation by an exchange of ions between the sample components and the loosely bound ions with the resins. Affinity Difference of sample analytes for the resins is the principle for separation.

- Size exclusion chromatography (SEC) is a form of LC wherein a uniform nonionic gel is used to separate sample components based on their molecular size. The small molecules get into the polymer network and are held back, whereas larger molecules cannot enter the polymer network and will be swept out of the column and separated earlier. The elution order is the largest molecules first, medium next and the smallest sized molecules last. It is also called as gel permeation chromatography (GPC) [18].

10.3.4 TYPE OF INFORMATION OBTAINED FROM CHROMATOGRAM

1. **Form:** The output of a chromatographic instrument can be of two types: a plot of retention time versus detector response. The peak area is a measure of the concentration of the component in the sample.

2. **Sample:**
   - Size: A few milligrams is usually enough for either GC or t/b.
   - State:
     - For GC, the sample can be gas, liquid, or solid. Solid samples are usually dissolved in a suitable solvent; both liquid or solid samples must volatilize at the operating temperature.
     - For LC, samples can be liquid or solid. Both must be soluble in the moving phase.

10.3.5 ADVANTAGES

1. **Gas Chromatography:**

   - Moderately fast quantitative analyses (0.5–1.5 hours per sample).
   - Excellent resolution of various organic compounds.

- Not limited by sample solubility.
- Good sensitivity.
- Specificity.

2. Liquid Chromatography:

- Separation of high boiling compounds.
- Not limited by sample volatility.
- Moving phase allows additional control over the separation.

10.3.6 LIMITATIONS

1. Gas Chromatography:

- Limited by sample volatility.

2. Liquid Chromatography:

- Less sensitive than GC.
- Detectors may respond to solvent carrier, as well as to sample.

3. Interferences: This in chromatography can generally be overcome by developing a right optimized method for separation.

10.3.7 TYPICAL HPLC CHROMATOGRAM

Figure 10.2 shows a typical chromatogram as an output of any chromatographic technique, which is basically a plot of detector signal as a function of time in minutes. The number above each peak (corresponds to a sample component) is retention time, which is a characteristic of an analyte separated during separation (elution). Peak area is a measure of the concentration of the component separated.

10.3.8 APPLICATIONS OF CHROMATOGRAPHY

1. Pharmaceuticals industry:

- Drug stability control.
- Quantitation of drug in the pharmaceutical formulation, e.g., Paracetamol in panadol tablet.

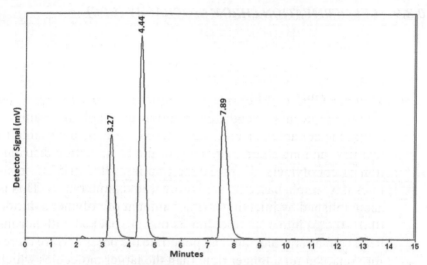

FIGURE 10.2 Typical chromatogram.

- Quantification of drug in biological fluids, e.g., blood glucose level.

2. **Analysis of natural contamination:**

- Phenol and mercury from seawater.

3. **Forensic test:**

- Analysis of steroid in blood, urine & sweat.
- Analysis of psychotropic drug in plasma.

4. **Clinical test:**

- Monitoring of hepatic cirrhosis patient through aquaporin 2 in the urine.

5. **Food and essence manufacture:**

- Analysis of sweetener in the fruit juice.
- Analysis of preservative in sausage.

10.3.9   GEL PERMEATION CHROMATOGRAPHY (GPC)

a. **Principle:** Separation of analytes in a sample based on the difference in molecular size using a polymer gel network as a separation medium.

b. **Theory:** GPC or SEC is a technique used to determine the average molecular weight distribution of a polymer sample. Using the suitable detectors and method, it is also possible to identify the long chain branching or the composition distribution of copolymers. As the name implies, GPC or SEC separates the sample based on size or hydrodynamic radius. This is accomplished by injecting a small amount of polymer solution (0.01–0.6%) into a set of columns that are packed with porous beads. Smaller molecules can penetrate the pores and are therefore retained for a longer time than the larger molecules which move through the polymer beads in the columns and elute faster. One or more detectors are attached to the output of the columns [19]. For routine analysis of linear homopolymers, often a Differential Refractive Index (DRI) or a UV detector. For branched or copolymers, however, it is necessary to have at least two sequential detectors to determine molecular weight accurately. Branched polymers can be analyzed using a DRI detector coupled with an on-line viscometer (VIS) or a low-angle laser light scattering (LALLS) detector. The compositional distribution (a function of molecular size) of copolymers can be determined using a DRI detector coupled with a UV or FTIR. It is important to consider the type of polymer and information to be obtained before submitting a sample (Figure 10.3).

10.3.9.1   GPC/DRI

Linear homopolymers, such as polyethylene, polystyrene, etc., can be analyzed using a single DRI detector. A calibration curve for a given reference polymer and the molecular weight is used to calculate the molecular weight of the sample polymer on knowing the Mark-Houwink constants, k, and α. For the branched polymer, require a secondary detector like LALLS or VIS for accurate results [20].

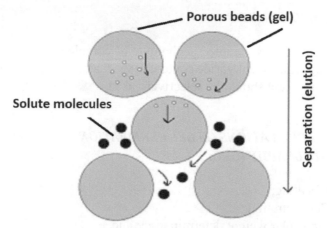

FIGURE 10.3 Working of gel permeation chromatography.

10.3.9.2 GPC/DRI/LALLS

Two detectors in tandem like LALLS and DRI can also be used. It is used to analyze PE, PP samples. The data comprises of two chromatograms corresponding two detectors. The DRI output gives the concentration profile while the LALLS gives more sensitivity at higher molecular weights.

10.3.9.3 GPC/DRI/VIS

GPC, coupled with an online VIS, can replace a LALLS detector to analyze branched-chain polymers. Here the intrinsic viscosity is measured even without Mark-Houwink constants and is complementary to the LALLS detector.

10.3.9.4 GPC/DRI/UV

The UV detector is used to analyze chromophores in graft or block copolymers. The molecular weight from the DRI calibration curve, and the effective extinction coefficient E' from the UV gives UV/DRI ratio. The GPC/DRI/UV combine can be employed to analyze polymer samples that are soluble in THF at 30–45°C.

10.3.9.5   GPC/DRI/FTIR

The GPC/DRI/FTIR coupled technique is complementary to the UV detector, e.g., maleic anhydride content in the maleated EP. This method is labor-intensive and should be selectively used.

10.3.10   APPLICATIONS OF GEL PERMEATION CHROMATOGRAPHY (GPC)

- Desalting;
- Concentration;
- Molecular weight determination; and
- Fractionation.

10.4   THERMAL ANALYSIS (TA)

1. **Principle:** Thermal analysis (TA) refers to a spectrum of techniques wherein **a** sample's thermal behavior is continuously quantified through a programmed temperature range.
2. **Theory:** The basic instrumental requirements are:

- A precision balance;
- A programmable furnace;
- A recorder.

TA is a very simple technique for quantitatively analyzing for filler content of a polymer compound (e.g., decomposition of carbon black in the air). The relationship between "structure" and "properties" of a polymer is the fallout of increasing molecular weight and is reflected upon its physical (mechanical) state; its journey from an oily liquid to a soft viscoelastic solid, to a hard, glassy elastic solid. Even slight atomic rearrangements can have dramatic effects, e.g., in the atactic and syndiotactic stereoisomers of polypropylene, the former behaves as a viscoelastic amorphous polymer at RT while the later one is a strong, fairly rigid plastic with a melting point above 160°C [21].

At high-temperature, conformational changes are frequent while at lower temperatures the chains solidify by either of two mechanisms:

ordered molecular packing in a crystal lattice, crystallization, or by gradual freezing out of long-range molecular motions, vitrification. These transformations, which define the principal mechanical behavior, the melt, the rubbery state, and the semicrystalline and glassy amorphous solids, accompany transitions in thermodynamic properties at the glass transition temperature, the melting, and the crystallization temperatures. TA techniques are designed to measure the above-mentioned changes both by measurements of heat capacity and stiffness [22].

Modern instruments are automated and interfaced with software for data acquisition. The samples are provided with a suitable environment with air, nitrogen, or oxygen as required. In TA, the mass loss can be due to events such as volatilization of liquids, decomposition, and evolution of gases from solids. The onset of volatilization is proportional to the boiling point of the liquid. The residue remaining at high temperature represents the percent ash content of the sample [23]. Among, the most common techniques are:

- Thermogravimetric analysis (TGA); and
- Differential scanning calorimetry (DSC).

10.4.1   THERMOGRAVIMETRIC ANALYSIS (TGA)

1. **Principle:** Mass losses due to different effects like desorption or decomposition with temperature is plotted as a thermogram.
2. **Theory:** TGA asks for a continuous weighing of a small sample (e.g., 10 mg) in a controlled atmosphere (e.g., air or nitrogen) as the temperature is increased at a given rate. The thermogram illustrates mass losses due to different effects like desorption of gases (e.g., moisture) or decomposition (e.g., HBr loss from halobutyl, or CO from calcium carbonate filler) [24].

10.4.2   DIFFERENTIAL SCANNING CALORIMETRY (DSC)

1. **Principle:** The DSC measures the power (heat energy per unit time) differential between a small weighed sample and a reference material with both the sample and reference are exposed to the same temperature program.

2. **Theory:** The DSC requires:

- Two cells equipped with thermocouples;
- A programmable furnace, recorder; and
- Gas controller.

Automation is even more extensive than in TA due to the more complicated nature of the instrumentation and calculations.

DSC is often used in conjunction with TA to determine if a reaction is endothermic, such as melting, vaporization, and sublimation, or exothermic, such as oxidative degradation [25].

The DSC measures the power (heat energy per unit time) of a small weighed sample of polymer (10 mg) in a sealed aluminum pan referenced to an empty pan in order to maintain a zero temperature differential by scanning during heating and cooling [26] (Figure 10.4).

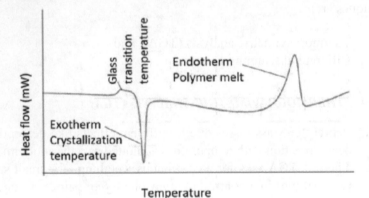

FIGURE 10.4   Thermogram.

10.4.3   *THERMOMECHANICAL ANALYSIS (TMA) OF POLYMERS*

1. **Principle:** A profile of thermal expansion coefficient or penetration distance versus temperature when the quartz probe is either rested or penetrated in the flat sample.

2. **Theory:** TMA consists of:

- A quartz probe which rests on top of a flat sample (a few mms square) in a temperature-controlled chamber. When setup in neutral buoyancy (with 'flat probe') then as the temperature is increased, the probe rises with the expansion of the sample giving thermal expansion coefficient versus temperature profiles.

- Alternatively, with the 'penetration probe' subjected to dead loading, a profile is obtained (penetration distance versus temperature). Although this is a simple and versatile experiment, it gives only a semi-quantitative indication of mechanical modulus versus temperature [27].

10.4.4 ADVANTAGES OF THERMAL ANALYSIS (TA)

- It is also used to determine the glass transition temperature of polymers.
- Liquids and solids can be analyzed by both methods of TA.
- The sample size is usually limited to 10–20 mg.
- TA can be used to characterize the physical and chemical properties of a system under conditions that simulate real-world applications.
- Used to measure the heat of fusion of polymers.
- Used to study the kinetics of chemical reactions, e.g., oxidation, and decomposition.
- Heat of fusion can be converted to degree of crystallinity provided, the heat of fusion for the 100% crystalline polymer is known.

10.4.5 LIMITATIONS OF THERMAL ANALYSIS (TA)

- Much of the data interpretation is empirical in nature.
- More than one thermal method may be required to fully understand the chemical and physical reactions occurring in a sample.
- Condensation of volatile reaction products on the sample support system of a TA can give rise to anomalous weight changes.

10.5 MICROSCOPY

1. **Principle:** Magnification and resolution of the samples under investigation using microscopic techniques for getting an idea about the particle size, morphology, etc.

2. Theory: Phase contrast—thin sections (100–200 nm) and refractive indices differing by approximately 0.005 are mounted on glass slides and examined as it is or with oil to remove microtoming artifacts, e.g., determination of the number of layers in coextruded films, dispersion of fillers and polymer domain size [28].

Polymer blends with a minimum domain size of 1 pm can be examined in the optical microscope using one or more of the following techniques:

- Polarized light is used if one of the polymer phases is crystalline or inorganic filters are agglomerated, e.g., nylon/EP blends and fillers such as talc.
- Incident is used to examine surfaces of bulk samples, e.g., carbon black dispersed in rubber.
- Bright field is used to examine thin sections of carbon black loaded samples, e.g., carbon black dispersion in thin films of rubber compounds.

When the domain size is in the range of < 1 μm to 10 nm, either SEM and/or TEM are required [29, 30].

Samples in the SEM can be examined either:

- for general morphology as it is.
- for freeze fractured surfaces.
- for microtome blocks of bulk solid samples.

Contrast is achieved by any one or combination of the following methods:

- **Solvent Etching**—when there exists a large solubility difference in a particular solvent of the polymers, e.g., PP/EP blends.
- **O_5O_4 Staining**—when there exists at least 5% unsaturation in the polymers, e.g., NR/EPDM, BIIR/Neoprene.
- **RuO_4 Staining**—when there exist either no solubility or unsaturation differences, e.g., dynamically vulcanized alloys [31, 32].

10.5.1 SCANNING ELECTRON MICROSCOPY (SEM)

SEM can be used:

- To study liquids or temperature-sensitive polymers on a cryostage.
- To do X-ray/elemental analysis.
- This technique is qualitative.
- X-ray analysis and mapping of the particular elements present is useful for the identification of inorganic fillers and their dispersion, inorganic impurities in gels or on surfaces and curatives, e.g., aluminum, silicon, or sulfur in rubber and Cl and Br in halobutyls [33].

10.5.2 TRANSMISSION ELECTRON MICROSCOPY (TEM)

TEM is used when:

- A more in-depth study (when domain sizes are < 1 μ or so) is required on polymer phase morphologies, e.g., dynamically vulcanized alloys, carbon black filler location in rubber compounds and also in the morphology of block copolymers.
- Thin sections are required and may take time, depending on the nature of the sample.
- They are prepared by microtoming with a diamond knife at near liquid nitrogen temperatures (–150°C).
- The contrasting media for TEM is same as for SEM.

10.5.3 NON-ROUTINE TECHNIQUES

The non-routine techniques are:

- Solvent casting when microtoming is not desirable as a sample preparation method.
- STEM – to compute elemental composition study in thin films when the better resolution is required than X-ray analysis in the SEM on bulk samples.
- Cryostage – SEM to examine liquid samples at low temperatures, e.g., butyl slurry.

- Fluorescence microscope – to study polymer/asphalt blends or any which is a fluorescent sample.
- OM/Hot stage – to observe the melting point of either an impurity or other moiety in a compound [34, 35].

10.6 ELEMENTAL AND STRUCTURAL CHARACTERIZATION

10.6.1 ATOMIC ABSORPTION SPECTROSCOPY

1. **Principle:** The quantitative estimation of the element of interest is determined based on the attenuation or absorption, of a characteristic wavelength emitted from a light source by the analyte atoms.
2. **Theory:** In atomic absorption spectrometry (AAS) the sample is vaporized and the element of interest atomized at high temperatures. The light source consists of a hollow cathode lamp containing the element to be measured. Separate lamps are required to examine each element. The dispesing element is used to isolate the required wavelength of light for the element in question, and the light source is modulated to reach only the monochromatic radiation not laced with any unwanted radiation to the detector. [36]

Conventional AAS instruments analyze liquid samples such as dilute acid and xylene solutions. The conventional AAS instruments use:

- A flame atomization system for liquid sample vaporization.
- An air-acetylene flame (2300°C) for most elements.
- Higher temperature nitrous oxide-acetylene flame (2900°C) for more refractory oxide-forming elements.
- Electrothermal atomization techniques like a graphite furnace can be used for the direct analysis of solid samples.

The sample volume needed is dependent on the number of elements to be analyzed. Flame atomization generates ions as well as atoms. Since only atoms are detected, the ratio of atoms/ions remains constant for a given element and is influenced by the presence of other elements in the sample matrix. The addition of large amounts of an easily ionized element such as potassium to both the sample and standards helps mask the ionization interference [37].

The capabilities of Flame AAS can be extended by employing the following changes:

1. A cold quartz tube for carrying mercury vapor for the determination of mercury;
2. A heated quartz tube for decomposing metallic hydride vapors for As, Se, Sb, Pb, Te, Sn, and Bi determination;
3. A graphite quartz tube for decomposing nonvolatile compounds of metals, with extremely high sensitivity.

Advantages of AAS:

- The AAS instruments are easy to operate.
- To determine metals at ppm levels.
- To analyze the light elements such as H, C, N, 0, P, and S, halogens, and noble gases.
- Higher concentrations can be determined by prior dilution of the sample.
- To determine boron and magnesium in oils.
- Offers excellent sensitivity for most elements with limited – interferences.
- Sensitivity can be extended up to sub-ppb range by flameless methods for some elements.

Limitations of AAS:

- AA is not recommended if a large number of elements are to be measured in a single sample.
- Although AA is a very capable, widely used technique worldwide, ICP, and XRF offer methods with a comparative advantage.
- The determination of several elements/sample is slow and requires larger sample size due to the sequential analytical method.
- Chemical and ionization interferences must be corrected by a correction in the volume of the sample solution, e.g., chemical interferences arise from the formation of thermally stable compounds such as oxides in the flame [38] (Figure 10.5).

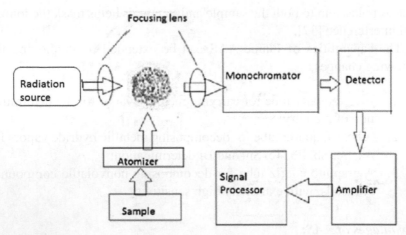

FIGURE 10.5   Block diagram of atomic absorption spectrophotometer.

10.6.2   *INDUCTIVELY COUPLED PLASMA* (ICP) *ATOMIC EMISSION SPECTROSCOPY*

1. **Principle:** The atoms of the sample are vaporized, atomized by collision with an extremely high-temperature plasma generated energetically excited argon species to emit characteristic spectra.
2. **Theory:** In inductively coupled plasma (ICP) atomic emission spectroscopy, the sample is vaporized and the element of interest atomized in an extremely high temperature (–7000°C) argon plasma, generated and maintained by radio frequency coupling. The atoms collide with energetically excited argon species and emit characteristic atomic and ionic spectra that are detected with a photomultiplier tube.

The separation of spectral lines can be accomplished in two ways:

a. In a sequential or scanning ICP, a scanning monochromator with a movable grating is used to bring the light from the wavelength of interest to a single detector.
b. In a simultaneous or direct reader ICP, a polychromator with a diffraction grating is used to disperse the light into its component wavelength. The solvent can be water, usually containing 10% acid, or a suitable organic solvent such as xylene [39].

Advantages of ICP:

- ICP offers good detection limits and a wide linear range for most elements.
- With a direct reading instrument, multi-element analysis is extremely fast.
- Since all samples are converted to simple aqueous or organic matrices prior to analysis, no reference is required.
- Chemical and ionization interferences frequently observed in AAS are suppressed in ICP analysis.
- With a scanning instrument, it may be possible to move to an interference-free line.

Limitations of ICP:

- ICP instruments are limited to the analysis of liquids only.
- Solid samples require some sort of dissolution procedure prior to analysis.
- The final volume of the solution for analysis should be at least 25 mL.
- Necessitates extensive sample preparation facilities and methods.
- ICP instruments are not rugged.
- Constant monitoring, especially of the sample introduction and torch systems, is essential.
- Spectral interferences can complicate the determination of trace elements in the presence of other major metals and must be corrected for accurate quantitative estimation [40].

Applications:

Sample types analyzed by ICP include:

- Trace elements in polymers, wear metals in oils, and numerous one-of-a-kind catalysts.

10.6.3   *ION CHROMATOGRAPHY (IC)*

1. **Principle:** Separation of ions of interest using ion-exchange resins by exchange reactions.

2. **Theory:** Commercial ion chromatography instruments have become available since early 1976. Ion chromatography (IC) is a combination of ion-exchange chromatography, eluent suppression, and conductimetric detection. For anion analysis, a low capacity anion exchange resin is used in the separator column, and a strong cation exchange resin in the H+ form is used in the suppressor column. A dilute mixture of Na_2CO_3 $NaHCO_3$, is used as the eluent, because carbonate and bicarbonate are conveniently neutralized to low conductivity species and the different combinations of carbonate-bicarbonate give variable buffered pH values. This allows the ions of interest in a large range of affinity to be separated. The anions are eluted through the separating column in the background of carbonate bicarbonate and conveniently detected based on electrical conductivity [41].

The reactions taking place on these two columns are, for an anion X:
A. **Separator Column:**
   $Resin - Na^+ HCO_3^- + Na^+X^- \rightleftharpoons Resin - Na^+X^- + Na^+HCO_3^-$
B. **Suppressor:**
   $Resin - SO_3^- + Na^+HCO_3^- \rightleftharpoons Resin - SO_3^-Na^+ + HCO_3^-$
   $Resin - SO_3^-H^+ + Na^+X^- \rightleftharpoons Resin - SO_3^-Na^+ + H^+X^-$

As a result of these reactions in the suppressor column, the sample ions are presented to the conductivity detector as H'X-, not in the highly conducting background of carbonate-bicarbonate, but in the low conducting background of HCO_3.

Dilute aqueous sample is injected at the head of the separator column. The anion exchange resin selectively causes the various sample anions of different types to migrate through the bed at different respective rates, thus effecting the separation. The effluent from the separator column then passes to the suppressor column where the H+ from cation exchange resin absorbs the cations in the eluent stream. Finally, the suppressor column effluent passes through a conductivity cell. The highly conductive anions in a low background conductance of HCO_3 are detected at high sensitivity by the conductivity detector. Because of the nonspecific nature of the conductivity detector, the chromatograph peaks are identified only by their retention times [42].

10.6.3.1   ADVANTAGES OF IC

A. Sequential multi-anion capability; eliminates individual determinations of anion by diverse technique.
B. Small sample size (< 1 mL).
C. Rapid analysis (–10 minutes for –7 anions).
D. Large dynamic range over four decades of concentration. E. Speciation can be determined.

10.6.3.2   DISADVANTAGES OF IC

A. Interferences possible if two anions have similar retention times.
B. Determination difficult in the presence of an ion present in very large excess over others.
C. Sample has to be in aqueous solution.
D. Method not suitable for anions with PKa of < 7.

In addition to the common inorganic anions analyzed by IC, a number of other species can also be determined by using appropriate accessories (Table 10.2).

TABLE 10.2   Applications of Ion Chromatography

Technique	Species
A. Ion chromatography	Carboxylic Acids
B. Chemistry – IC	Formaldehyde; Borate
C. Mobile Phase IC	Ammonia; Fatty Acids; Ethanol-Amines
D. Electrochemical Detection	Phenols, CN^-, Br^-, I^-, S^{-2}, etc.

10.6.4   MASS SPECTROSCOPY

1. **Principle:** Identification of sample analytes based on peaks of parent and fragment ions at given m/e ratio in the spectrum.
2. **Theory:** Most of the spectroscopic and physical methods employed by the chemist in structure determination are concerned only with the physics of molecules, mass spectroscopy deals with both the chemistry and the physics of molecules, particularly with gaseous ions. In conventional mass spectrometry (MS), the ions of interest are positively charged ions.

The mass spectrometer has three functions:

1. To produce ions from the molecules under investigation;
2. To separate these ions according to their mass to charge ratio; and
3. To measure the relative abundances of each ion.

Today, MS has achieved status as one of the primary spectroscopic methods, and the greatest advantage of the method is the extensive structural information which can be obtained from sub-microgram quantities of material.

The methodology of mass separation is governed by both the kinetic energy of the ion and the ion's trajectory in an electromagnetic field. There exists a balance between the centripetal and centrifugal forces which the ion experiences. Centripetal forces are caused by the kinetic energy and centrifugal forces by the electromagnetic field. This force balance is given as follows:

$$\frac{mU^2}{r} = qUB$$

where m = ion's mass U = ion's velocity, r = radius of ion trajectory in the magnetic field, q = ion's charge, B = magnetic field strength. The centripetal force (right-hand side), and the centrifugal force (left-hand side) is as shown in the above equation.

Solving for mass-to-charge ratio yields:

$$\frac{m}{q} = \frac{Br}{U}$$

The kinetic energy of the ion is given by:

$$aV = 1/2\ mu^2$$

where; q = charge of the ion; V = accelerating potential; m = ion's mass; and U = ion's velocity.

Solving for U yields:

$$\frac{2qV^{1/2}}{m}$$

By substituting for U in the second expression, we obtain:

$$m/q = Br/(2qV^{1/2}/m)$$

Squaring each side of the equation yields:

$$\frac{m}{q} = \frac{B^2}{2} \frac{e^2}{V}$$

Thus, the mass to charge ratio can be determined if one knows B, e, and V. Since e is constant for a mass passing through the two slivers, scanning a spectrum is achieved by varying either B or V, keeping the other constant.

The mass spectrum may be either in analog form (chart paper) or digital form (printed paper). Analyses are calculated to give mole%, weight%, or volume%. Either individual components, compound types by carbon number, or total compound type are reported [43].

The characteristics of the sample submitted for an MS test are:

A. **Size:** 1–1000 mg.
B. **State:** Gas, liquid, or solid, but only the portion vaporizable at about 300°C is analyzed.
C. **Phases:** If the sample has more than one phase, each phase can generally be analyzed separately.
D. **Composition Limitations:** Essentially, no limits to composition, simple mixtures, and complex mixtures can be handled.
E. **Temperature Range:** Samples should be at RT and should be thermally stable up to 300°C for bath introduction and maybe involatile for field desorption work.

10.6.4.1 ADVANTAGES

- Mass spectrometers provide a wealth of information about the structure of organic compounds, their elemental composition, and compound types in complex mixtures.
- A detailed interpretation of the mass spectrum frequently allows the positions of the functional groups to be determined.

- MS is used to investigate reaction mechanisms, kinetics, and is also used in tracer work.
- A wide variety of materials from gases to solids and from simple to complex mixtures can be analyzed.
- Only a very small amount of sample is required.
- The molecular weight and atomic composition are generally determined.

10.6.4.2 LIMITATIONS

- Some compounds such as long-chain esters and polyethers decompose in the inlet system, and the spectrum obtained is not that of the initial substance.
- Calibration coefficients are required for quantitative analyses.
- The sample introduced to the instrument cannot usually be recovered.
- Some compounds, such as olefins and naphthenes, give very similar spectra and cannot be distinguished except by analysis before and after hydrogenation or dehydrogenation.

10.6.5 NUCLEAR MAGNETIC RESONANCE (NMR) SPECTROSCOPY

1. **Principle:** Determination of chemical structure depending upon the energy required for the alignment behavior of atomic nuclei with a magnetic moment governed by magnetic field strength and chemical environment of the nucleus.

2. **Theory:** NMR is a spectrometric technique for determining chemical structures. When an atomic nucleus with a magnetic moment is placed in a magnetic field, it tends to align with the applied field. The energy required to reverse this alignment depends on the strength of the magnetic field and to a minor extent on the environment of the nucleus, i.e., the nature of the chemical bonds between the atom of interest and its immediate vicinity in the molecule. This reversal is a resonant process and occurs only under select conditions. By determining the energy levels of transition for all of the atoms in a molecule, it is possible to determine many important features of its structure. The energy levels can be expressed in terms of frequency of electromagnetic radiation, and

typically fall in the range of 5–600 MHz for high magnetic fields. The minor spectral shifts due to the chemical environment are the essential features for interpreting structure and are normally expressed in terms of part-per-million shifts from the reference frequency of a standard such as tetramethylsilane [44].

The most common nuclei examined by NMR are ¹H and ¹³C, as these are the NMR sensitive nuclei of the most abundant elements in organic materials. ¹H represents over 99% of all hydrogen atoms, while ¹³C is only just over 1% of all carbon atoms. Until fairly recently, instruments did not have sufficient sensitivity for routine ¹³C NMR, and ¹H was the only practical technique.

In general, the resonant frequencies can be used to determine molecular structures. 'H resonances are fairly specific for the types of carbon they are attached to, and to a lesser extent to the adjacent carbons. These resonances may be split into multiples, as hydrogen nuclei can couple to other nearby hydrogen nuclei. The magnitude of the splitting and the multiplicity can be used to better determine the chemical structure in the vicinity of a given hydrogen. Complications can arise if the molecule is very complex, because then the resonances can overlap severely and become difficult or impossible to resolve [45].

¹³C resonances can be used to directly determine the skeleton of an organic molecule. The resonance lines are narrow, and the chemical shift range (in ppm) is much larger than for ¹H resonances. Furthermore, the shift is dependent on the structure of the molecule for up to three bonds in all directions from the site of interest. Therefore, each shift becomes quite specific, and the structure can be easily assigned, frequently without any ambiguity, even for complex molecules.

Very commonly, however, the sample of interest is not a pure compound, but is a complex mixture such as a coal liquid. As a result, a specific structure determination for each molecular type is not practical, although it is possible to determine an average chemical structure. More details are possible, but depend greatly on the nature of the sample, and what information is desired.

Any gas-liquid or solid sample that can be dissolved in solvents, such as CCl, CH – Cl, acetone, or DMSO to the one percent level or greater can be analyzed by this technique. Samples of –0.1 g or larger of pure material are sufficient. Solids can also be analyzed as solid. However, special arrangements need to be made. In either case, the analysis is non-destructive so that samples can be recovered for further analysis if necessary [46].

The NMR experiment can be conducted in a temperature range from liquid nitrogen (−209°C) to +150°C. This gives the experimenter the ability to slow down rapid molecular motions to observable rates or to speed up very slow or viscous motions to measurable rates.

10.6.5.1 ADVANTAGES

- NMR is a very powerful tool.
- It often provides the best characterization of compound structure.
- It may provide absolute identification of specific isomers in simple mixtures.
- It may also provide a general characterization by functional groups which cannot be obtained by any other technique [47].
- Interfaced with other techniques (such as mass or infrared spectrometry) NMR can often provide greatly improved characterizations.
- NMR results are quantitative.

10.6.5.2 LIMITATIONS

- Analysis of a ^{13}C or ^{1}H spectrum, would reveal the different types of functionalities, as well as their contents in the sample [48].

10.6.6 FOURIER TRANSFORM INFRARED SPECTROSCOPY (FTIR)

1. **Principle:** Structural information can be ascertained based upon the characteristic vibrational frequencies absorbed by the molecule which depends upon vibrational transitions it undergoes.

2. **Theory:** Fourier Transform Infrared Spectrometry (FTIR) is a special technique. It is used to analyze samples that are available either in small quantity or a small entity. Gels within a rubber sample, have to be microtomed (i.e., cut into very thin slices) and mounted on KBr plates in a Microscopy Lab. Samples contaminated with inorganic components are usually analyzed by both

X-ray and FTIR microscope. Sample size –20 microns can be analyzed by the FTIR-microscope. FTIR is commonly used for qualitative identification of various functionalities. For quantitative analysis, FTIR requires the use of well-characterized standards. In some cases, the assigned peak absorbance is relatively small, so a thick film, 6 mm, is used.

Vibrational transitions (mid-IR) are the most useful of all to study. These give information about the presence or absence of specific functional groups in a sample. Practically all functional groups (infrared-active) display that fundamental over a very narrow range of wavelength in the mid-infrared region. Moreover, the whole spectrum, containing fundamentals, overtones, and combination bands, constitutes a fingerprint of the sample. This means that although we might not know what a sample is, we will always know it later if it occurs again. Finally, the absorption intensity of any band, whether fundamental or overtone, is proportional to the number of functional groups giving rise to the signal [49].

Modern dispersive spectrometers are double beam ones which divide the incident beam into two; one beam goes through the sample, and the other goes through a suitable reference. The intensity of both beams are monitored by a suitable detector, and final data output can be displayed in either transmittance or absorbance:

$$\text{Transmittance} = 1_t/I_0$$

$$\text{Absorbance} = -\log 1_0/I_t$$

where I_S, I_R, I_t, I_0 refer to the intensities in the sample, reference, transmitted, and incident beam, respectively.

The characteristics of the sample used are as follows:

A. For gases, we generally need about 250 cm^3 at 1 atm to obtain a spectrum.
B. For liquids, we generally need about 0.25 cm^3 to obtain a spectrum.
C. For solids, we generally need about 1 mg to obtain a spectrum.
D. Trace analyses within samples will, of course, increase the sample requirements proportionally.

The advantages of this technique are:

A. Faster and cheaper than most other techniques.
B. Very specific for certain functional groups.
C. Very sensitive for certain functional groups.
D. Fingerprint capability.

The disadvantages of this technique are:

A. Requires special cells, NaCl, KBr, quartz, etc.
B. Usually requires solubility of sample.
C. Very difficult to get good quantitation in solids.
D. Must calibrate all signals.
E. Water interferes.

The measurement interferences can occur from:

A. Water interferes with practically all IR work.
B. Solvents generally interfere and must be selected carefully.
C. Multicomponent samples generally have mutually interfering species. Separations are often required. Sometimes, changing the spectral region helps.
D. Optical components interfere to different extents in different regions. Thus, quartz is good for UV/Vis/Near-IR, but bad for mid-IR/far-IR. KEIr is good for mid-IR, bad for far-IR [50].

10.6.7 *ULTRAVIOLET, VISIBLE, AND SPECTROSCOPY*

1. **Principle:** The advantage of absorption characteristics of a material when it is incident upon by UV-Visible electromagnetic radiation for unraveling the structural information.
2. **Theory:** When electromagnetic radiation passes through a sample, some wavelengths are absorbed by the molecules of the sample, and are said to be promoted to an excited energy state. The total energy state of the ensemble of molecules may be regarded as the sum of the four kinds of energy: electronic, vibrational, rotational, and translational energy.

The electronic absorptions are accompanied by changes in both rotational and vibrational properties of the molecules on the electronic transitions. Another complication arises in the interpretation of absorption spectra. If a molecule vibrates with pure harmonic motion and the dipole moment is a linear function of the displacement, then the absorption spectrum will consist of fundamental transitions only. If either of these conditions is not met, as is usually the case, the spectrum will contain overtones (multiples of the fundamental) and combination bands (sums and differences). Most of these overtones and combination bands occur in the near-infrared (0.8–20 μm). Not all vibrations and rotations are infrared-active. If there is no change in dipole moment, then there is no oscillating electric field in the motion, and there is no mechanism by which absorption of electromagnetic radiation can take place. Table 10.3 shows five important areas of the electromagnetic spectrum [51, 52].

TABLE 10.3 Regions of the Electromagnetic Spectrum

Wavelength Range (m)	Electromagnetic Radiation
0.2–0.4	Ultraviolet (electronic)
0.4–0.8	Visible (electronic)
0.8–2.0	Near-IR (overtones)
2.0–25.0	Mid-IR (vibrational)
25.0–500.0	Far-IR (rotational)

Electronic transitions (UV, visible spectra) generally give information about unsaturated groups in the sample molecules (Table 10.4).

TABLE 10.4 Characteristic Absorption Maximum for Some Classes of Compounds

Class of Compounds	Wavelength Range of Absorption (μm)
Olefins	0.22
Aromatics	0.26–0.28
Carbonyls	0.20–0.27
Poly-nuclear aromatics	0.26–0.50
Conjugated C=S groups	0.62
Any colored substance	Visible region

The intensity of the absorption is proportional to the number of chromophores giving rise to the absorption band.

3. **Applications:**

- Structural information like aromaticity or aliphaticity determination.
- Presence or absence of functional groups.
- Presence of colored or colorless compounds.
- Presence of extended conjugation.

10.7 NEUTRON ACTIVATION ANALYSIS

1. **Principle:** Radioactivity level of an element is measured by converting from nonradioactive to radioactive by high-energy neutron bombardment.

2. **Theory:** Neutron activation analysis is a method of elemental analysis in which nonradioactive elements are converted to radioactive ones by neutron bombardment, and the elements of interest are determined from resulting radioactivity. High energy ($\approx$14 MeV) neutrons are generated by the reaction of medium energy deuterium ions with tritium, e.g., for oxygen analysis, the carefully weighed sample is irradiated for 15 seconds to convert a small amount of the oxygen-16 to nitrogen-16, which emits gamma rays with a half-life of 7.4 seconds. The irradiated sample is transferred to a scintillation detector where the gamma rays are counted for 30 seconds to ensure that all usable radioactivities have been counted. The system is calibrated with standards of known oxygen content. The raw data in the form of counts/30 seconds from a digital counter can be converted to weight% of the element of interest [53, 54].

The characteristics of a proper sample are:
- Size – Container (plastic cylinder 9 mm I.D. x 20 mm deep. Holds –1. Cc).
- State and phases – Solid or liquid, reasonably homogeneous.
- Moisture – free sample if true sample oxygen is desired.
- Temperature range – RT only.

10.6.7.1   ADVANTAGES

- Principle method for determining total oxygen directly Concentration of oxygen, which can be determined (0.01–60%).
- Fast (about 10 minutes/analysis). Only 1 minute requires for replicate measurement.
- Nondestructive.
- Moderate sensitivity.

10.8   POLYMER SURFACE CHARACTERIZATION

1. **Principle:** Surface mapping techniques like XPS, ESCA, etc. are used for the elucidation of compositional and/or structural information from the spectra obtained.
2. **Theory:** In many applications, the surface properties of polymeric materials undermine the benefits derived from bulk properties. Inadequate adhesion is a notable example, which has led to surface pretreatment to modify surface chemistry. The region of structure-property determination generally extends only a few molecular layers from the surface into the solid. In order to investigate the surface behavior, it is necessary to obtain compositional and structural information, particularly from this depth of penetration [55, 56].

Two established techniques have been in use for the characterization of polymer surfaces, namely:

- X-ray photoelectron spectroscopy (XPS or ESCA); and
- Secondary ion MS operated in the 'static' mode (static SIMS).

The techniques are highly complementary and are increasingly used in conjunction.

3. **Application:**

- Determination of structural information of the surfaces;
- Determination of compositional information of the surfaces;
- Ascertaining surface topographical information (surface finish, texture, etc.).

10.8.1   X-RAY FLUORESCENCE SPECTROMETRY

1. **Principle:** The binding energy of photoelectrons ejected electrons by the sample on irradiating it with X-rays is measured for ascertaining the information about chemical composition etc.
2. **Theory:** Nowadays, XPS is considered a modern tool for carbonaceous materials characterization used to determine the chemical composition, impurity presence, and nature of chemical bonds. The XPS analysis performed on the CNT surface was focused on measuring the binding energy of photoelectrons ejected when CNTs are irradiated with X-rays [57, 58].
3. **Applications:**
    The most important application of XPS in NCs based on polymer-CNT is:

    *   To identify the chemical composition of the modified carbon nanotubes surface in order to confirm if the surface modification has occurred or not.
    *   To identify the presence of impurity.
    *   Type of chemical bonding.

10.9   BIODEGRADABILITY CHARACTERIZATION

1. **Principle:** Measurement of biodegradability of polymeric materials under the influence of microorganisms.
2. **Theory:** Biodegradable polymers and polymer NCs can be exposed to a wide variety of natural, simulated, or artificial environments to measure or monitor their extent of degradation. These environments can be used to assess the biodegradation behavior of the materials containing biodegradable polymers. These environments typically include soil, compost, marine substrates (sludge, water), and sewage. Each environment contains different microorganisms (species diversity and population), different physical and chemical parameters (temperature, moisture, pH, aeration, and nutrients) that influence rates of microbial activity which in turn reflect upon the rate of biodegradation. Consequently, the biodegradation should always be assessed in their intended environment of "end usage" as different environments are not always comparable [59–62].

10.9.1   *METHODS OF MEASURING BIODEGRADATION*

An ideal biodegradation test method should include conditions expected during the period of application and the disposal environment at the end of their life cycle. A host of the standard on-field and lab-based test methods are currently available to measure the biodegradability of polymers and polymer NCs [63].

Following are the test methods generally undertaken for the assessment of biodegradation.

10.9.2   *ANALYTICAL TECHNIQUES*

10.9.2.1   *MORPHOLOGICAL ANALYSIS*

Physical appearance of polymers (color, shape, size, any visible holes or cracks on the polymer surface, and/or biofilm formation) is recorded before, and after biodegradation, e.g., a colorimeter is commonly used for color measurements (yellowness, whiteness, light transmission, and haze) [64].

10.9.2.2   *MICROSCOPIC ANALYSIS*

Optical microscopy (OM) and SEM techniques are commonly used to investigate surface morphology, and to detect the growth of microorganisms on polymers during biodegradation while TEM is primarily used for elucidation of internal structure, and spatial distribution of the various phases in the polymer [65].

10.9.2.3   *GRAVIMETRIC ANALYSIS*

Polymer samples are weighed before and after biodegradation, and the percentage of weight loss is determined as follows:

$$\% \text{ weight loss} = \frac{W_{t0} - W_{ts}}{W_{t0}} \times 100$$

where, W_{to} and W_{ts} refer to the weight of samples at time 0 (before exposure) and at specific sampling time, respectively [66].

10.9.2.4   PHYSICAL AND THERMAL ANALYSIS (TA)

The host of physical and thermal analytical tests is carried out for the given kind of characterization as given in Table 10.5.

TABLE 10.5   Choice of Characterization Tools

Type of test	Kind of characterization
Tensile tests (strength, modulus, and elongation at break)	Variation in mechanical behavior
X-ray diffraction (XRD)/ wide-angle x-ray diffraction (WAXRD)	% crystallinity and morphology
Differential scanning calorimetry (DSC) and thermogravimetric analysis (TGA)	Thermal behavior of polymers (glass transition, softening point temperatures)
Dynamic mechanical thermal analysis (DMTA) tests	Response to cyclic deformation with variation in the temperature

10.9.2.5   SPECTROSCOPIC ANALYSIS

Fluorescence and UV-visible spectrometry are used to determine the transmission of light through polymer films. Transparency to UV and visible light measurements are important for choosing the right packaging materials to preserve the quality of packaged food products [88]. FTIR, NMR, and MS are routinely used to obtain qualitative information on chemical structure or functional groups in polymers during biodegradation. Hyphenated methods are used to gain better insights into the degree and distribution of nanoscale fillers embedded within the body of polymer matrix [67, 68].

10.9.2.6   CHROMATOGRAPHIC ANALYSIS

GPC is often used to determine changes in average molecular weights and polydispersity index (Mw/Mn) of polymeric materials during biodegradation. High-performance liquid chromatography (HPLC) and GC techniques are used to detect monomers and oligomers formed during biodegradation [69].

10.9.2.7 RESPIROMETRIC ANALYSIS

Indirect evaluation of biodegradation can be done by more than one method. These methods are based upon

- The quantification of CO_2 evolved or;
- Oxygen consumed discretely or continuously on aerobic or anaerobic digestion of the material and then;
- Correlated with the extent of biodegradation.

$$\text{Aerobic: } C_{Sample + O_2} = CO_2 + C_{Biomass} + C_{Residual}$$

$$\text{Anaerobic: } C_{Sample + H_2} = CO_2 + CH_4 \, C_{Biomass} + C_{Residual}$$

Some samples (i.e., dumbbells, and other simple shapes) can be simultaneously tested for visual or physical changes, such as weight loss, mechanical or electrical testing. However, such tests must be dealt with caution while either interpolating or extrapolating the data because of the complex nature of interactions and the possibility of residual contamination from the medium [70].

The evolved gases can be analyzed by GC directly, infrared spectroscopy, or absorbed into alkaline solutions (such as $Ba(OH)_2$ or $NaOH$) for volumetric estimations. The GC chromatogram of the sample against the reference can give an idea about the extent of degradation. The volumetric estimation is calculated as:

$$\textbf{\textit{Sample degradation}}(\%) = \frac{Vol(CO_{2 \, SAMPLE+MEDIUM}) - Vol(CO_{2 \, MEDIUM})}{Vol(CO_{2 \, MEDIUM})}$$

Similarly, oxygen demand too can be a measure of the extent of biogradability of the sample in question [71].

10.9.3 MICROBIOLOGICAL TECHNIQUES

Microbial degradation of materials is assessed either by subjecting the test materials to a host of microorganisms from compost, soil, marine, freshwater, and activated sludge or selected pure cultures of bacteria or fungi with known biodegradation potential [72].

10.9.3.1 DIRECT CELL COUNT TECHNIQUE

Bacterial count grown up on the surface of test materials can be determined by epifluorescence microscopy using special kits.

10.9.3.2 CLEAR – ZONE TECHNIQUE

Test material is placed on a solid medium in a Petri dish containing no additional C source and sprayed on with a mixed inoculum of known bacteria and fungi. After incubation at specified test conditions (30–50°C; 3–4 weeks), the test material is examined for either surface growth and/ or formation of the clear zone (inhibition zone) showing susceptibility or resistance to microbial degradation of it [73, 74].

10.9.3.3 POUR PLATE/STREAK PLATE TECHNIQUE

This technique is used frequently for the enumeration and isolation of biodegrading microorganisms. Samples are taken from the surface of degraded material (previously exposed to environmental or pure cultures), and serial dilutions are made with a sterile solution. Pour plate method involves adding an aliquot of the dilution into a Petri plate, adding known quantity of agar medium, mixing, and incubating at test conditions conducive to the growth of microorganisms while in the streak plate method, an aliquot is added to agar medium in a Petri plate, microbial cells spread across the plate with either a wire loop or a spreader to manually isolate the cells such that each cell grows into a separate colony during incubation [75–77].

10.9.3.4   TURBIDITY DETERMINATION

Bacterial growth on the biodegraded test materials can be estimated using a spectrophotometer. Bacterial cells absorb or scatter light, given their cell mass or number. An increase in the turbidity of liquid media shows an increase in microbial cell count capable of utilizing the carbon in the test material [78, 79].

10.9.4   ENZYMATIC TECHNIQUES

Enzymes catalyze the breakdown of specific bonds in polymers forming monomers and oligomers, resulting into gravimetric and molecular weight loss. It is carried out under a defined set of laboratory conditions. Enzymatic activity is measured spectrophotometrically, and the% weight loss over the time is calculated [80].

10.9.5   MOLECULAR TECHNIQUES

Molecular techniques are considered powerful diagnostic and quantitative tools for the study of microorganisms in their natural habitats. These methods target functional or phylogenetically informative genes, or RNA to derive genetic information which reflects upon a microbial structure, and diversity to ecosystem functions [81].

10.10   CONCLUSION

Polymers and polymer-based products have to be analyzed both qualitatively and quantitatively in order to assure their quality and utility. There are numerous wet chemistry and instrumental based or classical and non-classical techniques either in isolation or coupled with other techniques have been in use to characterize the polymeric materials. The choice of a right technique is a must as every analytical tool diagnoses a property or two of the material in question. To get a comprehensive view and a better insight into the material identification, the right set of tools is a must. Preparation of the sample (representative of the bulk), sample preparation and in a given form is a prerequisite for accuracy in the measurements and thereby ascertaining right information.

Polymers can be analyzed by using classical chemical analysis methods which involve gravimetric and volumetric analysis, chromatography, etc. These methods have their own advantages and limitations, e.g., large amount solvents and other auxiliary agents, time-consuming, loss of material or destructive sample analysis methods, more sample size for analysis, lack of reproducibility, etc.

Given the limitations of classical methods more advanced, sophisticated, automated, software interfaced instrumental based methods have been in use due to their comparative advantage over wet chemistry-based techniques in terms rapid analysis, a substantial reduction in sample size, higher degree of precision and accuracy, reproducibility, online analysis, etc.

Chromatographic analysis is an important tool wherein the sample to be investigated is made into a solution with a right kind of solvents and by developing a right method sample analytes are analyzed say in HPLC. These techniques help analyze thermally labile, molecular, polar, or nonpolar materials with a great degree of accuracy and reproducibility. There has been an evolution in HPLC with the development of advanced techniques like UPLC, UFLC, etc., which further improve sensitivity, selectivity, rapidity, etc., in the analysis. GC requires samples to be analyzed, which are necessarily volatile and thermally stable, and by developing a right method based on the boiling range of the components, analysis can be done accurately and reproducibly both qualitative and quantitative. IC is also a type of LC wherein the ion-selective membranes find use in analyzing ionic species in the sample under investigation. Hyphenated techniques like LC-MS or GC-MS etc. are required to be used when LC or GC in isolation is unable to qualitatively and or quantitatively analyze the analytes in question. GPC helps in measuring molecular weight and its distribution profile as polydispersity index from the data generated for the polymers. There has been a lot of evolution in the GPC systems which are provided with detectors like DRI, UV-Vis, FTIR, LALLS, etc. which are used depending upon the requirements of the sample in the question of the polymer.

TA techniques such as TGA, DSC, TMA are put into use to follow the kinetics and thermodynamics of the reactions like decompositions, oxidation, etc. The data generated from the thermograms help in determining parameters like glass transition temperature, the heat of fusion, % crystallinity, etc.

Optical and electron microscopic examination tools such as SEM, TEM, and AFM, etc. are used to characterize polymers in order to get an insight into them. The high-resolution microscopy gives an idea about the particle size and morphology and position or location of components in a sample with high resolution and magnification power.

Spectroscopy is also a most sought after analytical tool to get an understanding of the structural features like aromaticity or aliphaticity, degree of saturation and or unsaturation, functional groups, presence or absence of heteroatoms, nature, and number of protons etc. using the absorption characteristics of different molecules or atoms of elements in the polymer samples by way of tools like UV-Vis, FTIR, NMR, AAS, ICP, NAA etc. Both qualitative and quantitative analysis with a great degree of accuracy is possible on developing calibration curves with suitable references. When it comes to surface characterization for the functionality of the polymers, surface topography, and elemental composition, X-ray fluorescence spectroscopy is an important tool used on priority. There are numerous techniques that can be used to characterize the biodegradation of polymers and NCs right from simple gravimetric analysis to advanced analytical techniques. Biodegradability of polymers can be tested by using physical, gravimetric, spectroscopic, chromatographic, enzymatic, microbiological, respirometric, etc. testing methods. Both techniques and nature of the environments in which polymers and NCs can be exposed, include variety in terms of natural, simulated, or artificial ones. These are like soil, compost, sludge, water, and sewage, etc. Each environment is laden with varied microorganisms in the form of species, their diversity, population, physical factors (like temperature, moisture, pH, aeration) that affect the pace of microbial action and in turn the rate of degradation observed in the polymeric materials. Consequently, the biodegradation of polymeric materials has to be assessed in their intended practical environment give their end-use. Neutron activation analysis method finds use in elemental analysis of a sample [82].

KEYWORDS

- atomic force microscopy
- continuous wave and pulsed ESR spectroscopy
- dielectric relaxation spectroscopy
- differential scanning calorimetry
- electron spin resonance
- Fourier transform infrared spectroscopy
- nuclear magnetic resonance

- Raman spectroscopy
- scanning electron microscopy
- thermogravimetric analysis
- transmission electron microscopy
- x-ray diffraction
- x-ray photoelectron spectroscopy

REFERENCES

1. Lan, T., Kaviratna, P. D., & Pinnavaia, T. J., (1994). *Chem. Mater., 6,* 573.
2. Chin, I. J., Thurn, A. T., Kim, H. C., Russell, T. P., & Wang, J., (2001). *Polymer, 42,* 5947.
3. Lim, S. K., Kim, J. W., Chin, I. J., Kwon, Y. K., & Choi, H. J., (2002). *Chem. Mater., 14,* 1989.
4. Wang, Z., & Pinnavaia, T. J., (1998). *Chem. Mater., 10,* 3769.
5. Messersmith, P. B., & Giannelis, E. P., (1995). *J. Polym. Sci., 33,* 1047.
6. Yano, K., Usuki, A., & Okada, A., (1997). *J. Polym. Sci., 35,* 2289.
7. Shi, H., Lan, T., & Pinnavaia, T. J., (1996). *Chem. Mater., 8,* 1584.
8. Giannelis, E. P., (1996). *Adv. Mater., 8,* 29.
9. LeBaron, P. C., Wang, Z., & Pinnavaia, T. J., (1999). *Appl. Clay Sci., 15,* 11.
10. Bailey, S. W., (1984). *Reviews in Mineralogy* (Vol. 157, pp. 101–120). Virginia Polytechnic Institute and State University, Blacksburg.
11. Brindley, G. W., & Brown, G., (1980). *Crystal Structures of Clay Minerals and Their X-Ray Identification.* Mineralogical Society, London.
12. Theng, B. K. G., (1974). *The Chemistry of Clay – Organic Reactions.* John Wiley & Sons, Inc., New York.
13. Jasmund, K., & Lagaly, G., (1993). *Tonminerale und Tone Struktur.* Steinkopff, Darmstadt.
14. Osman, M. A., Ploetze, M., & Suter, U. W., (2003). *J. Mater. Chem., 13,* 2359.
15. Pinnavaia, T. J., (1983). *Science, 220,* 365.
16. Krishnamoorti, R., & Giannelis, E. P., (1997). *Macromolecules, 30,* 4097.

17. Osman, M. A., Mittal, V., Morbidelli, M., & Suter, U. W., (2003). *Macromolecules, 36*, 9851.
18. Mittal, V., (2008). *J. Mater. Sci., 43*, 4972.
19. Batchelder, D. N., (1988). *Euro. Spectrosc. News, 80*, 28.
20. Young, R. J., Lu, D., & Day, R. J., (1991). *Polym. Int., 24*, 71.
21. Zimba, E, G., Hallmark, V. M., Swalen, J. D., & Radbolt, J. F., (1987). *Appl. Spec., 41*, 721.
22. Purcell, F. J., (1990). *Spectrosc. Int. 1, 33,*
23. Purcell, F. J., (1989). *Spectroscopy, 4*, 24.
24. Batchelder, D. N., Cheng, E., & Pitt, G. D., (1991). *Advanced Materials, 3*, 566.
25. Mitra, V. K., Risen, Jr., W. M., & Baughman, R. H., (1977). *J. Chem. Phys., 66*, 2731.
26. Batchelder, D. N., & Bloor, D., (1979). *J. Polym. Sci., Polym. Phys. Ed., 17*, 569.
27. Galiotis, E., Young, R. J., & Batchelder, D. N., (1983). *J. Polym. Sci. Polym. Phys. Edn., 21*, 2483.
28. Batchelder, D. N., & Bloor, D., (1984). 'Resonance Raman spectroscopy of conjugated macromolecules.' In: Clark, R. J. H., & Hester, R. E., (eds.), *Advances in Infrared and Raman Spectroscopy* (Vol. 11, 150–160). Wiley Heyden Ltd., Chichester.
29. Galiotis, E., (1982). *Polydiacetylene Single Crystal Fibers,* PhD Thesis. University of London.
30. Wu, G., Tashiro, K., & Kobayashi, M., (1989). *Macromolecules, 22*, 188.
31. Young, R. J., (1987). 'Polymer single crystal fibers.' In: Ward, I. M., (ed.), *Developments in Oriented Polymers* (Vol. 3, pp. 96–108). Applied Science, London.
32. Wegner, G., (1977). *Pure Appl. Chem., 49*, 443.
33. Wegner, G. Z., (1969). *Naturforsch., 24b*, 824.
34. Bloor, D., Ando, D. J., Preston, F. H., & Stevens, G. C., (1974). *Chem. Phys. Lett., 24*, 407.
35. Bloor, D., Koski, L., Stevens, G. C., et al., (1975). *Mater. Sci., 10*, 1678.
36. Galiotis, C., & Young, R. J., (1983). *Polymer, 24*, 1023.
37. Galiotis, C., Read, R. T., Yeung, P. H. J., et al., (1984). *Polym. Sci. Polym. Phys. Edn., 22*, 1589.
38. Lewis, W. F., & Batchelder, D. N., (1979). *Chem. Phys. Lett., 60*, 232.
39. Cottle, A. C., Lewis, W. F., & Batchelder, D. N., (1978). *J. Phys. C, 11*, 605.
40. Treloar, L. R. G., (1960). *Polymer, 1*, 95.
41. Baughman, R. H., Gleiter, H., & Sendfeld, N., (1974). *J. Polym. Sci. Polym. Phys. Edn., 12*, 1511.
42. Capaccio, G., & Ward, I. M., (1974). *Polymer, 15*, 233.
43. Capaccio, G., Gibson, A. G., & Ward, I. M., (1979). In: Herri, A. C., & Ward, I. M., (eds.), *Ultra-High Modulus Polymers*. Applied Science, London.
44. Zachariades, A. E., Mead, W. T., & Porter, R. S., (1979). In: Herri, A. C., & Ward, I. M., (eds.), *Ultra-High Modulus Polymers* (Vol. 1, pp. 5–9). Applied Science, London.
45. Smith, P., & Lemstra, P. J., (1979). *Makromol. Chem., 180*, 2983.
46. Smith, P., & Lemstra, P. J., (1981). *J. Mater. Sci., 15*, 505.
47. Holliday, L., & White, J. W., (1971). *Pure Appl. Chem., 26*, 545.
48. Kinloch, A. J., & Young, R. J., (1983). *Fracture Behavior of Polymers*. Applied Science, London.
49. Sakurada, I., Ito, K., & Nakamae, K., (1966). *J. Polym. Sci. CIS, 75*.

50. Clements, J., Jakeways, R., & Ward, I. M., (1978). *Polymer, 19*, 639.
51. Nakamae, K., Nishino, T., & Ohkubo, H., (1991). *J. Macromol. Sci. Phys., 830*, 1.
52. Grubb, D. T., & Liu, J. J. H., (1985). *J. Appl. Phys., 58*, 2822.
53. Prasad, K., & Grubb, D. T., (1990). *J. Polym. Sci.: Part B: Polym. Phys., 28*, 2199.
54. Roylance, D. K., & Devries, K. L., (1971). *J. Polym. Sci., Polym. Lett., 9*, 443.
55. Wool, R. P., & Statton, W. O., (1974). *J. Polym. Sci., Polym. Phys. Edn., 12*, 1575.
56. Wool, R. P., Bretzlaff, R. S., Li, B. Y., et al., (1986). *J. Polym. Sci. Polym. Phys. Edn., 24*, 1039.
57. Tashiro, K., Wu, G., & Kobayashi, M., (1988). *Polymer, 29*, 1768.
58. Jayasekara, R., Lonergan, G. T., Harding, I., Bowater, I., Halley, P., & Christie, G. B., (2001). An automated multi – unit composting facility for biodegradability evaluations. *J. Chem. Technol. Biotechnol., 76*, 411–417.
59. Kijchavengkul, T., Auras, R., Rubino, M., Ngouajio, M., & Thomas, F. R., (2006). Development of an automatic laboratory – scale respirometric system to measure polymer biodegradability. *Polym. Test., 25*, 1006–1016.
60. Kale, G., Auras, R., Singh, S. P., & Narayan, R., (2007). Biodegradability of polylactide bottles in real and simulated composting conditions. *Polym. Test., 26*, 1049–1061.
61. Calmon, A., Dusserre, B. L., Bellon, M. V., Feuilloley, P., & Silvestre, F., (2000). An automated test for measuring polymer biodegradation. *Chemosphere, 41*, 645–651.
62. Drímal, P., Hoffmann, J., & Druzbík, M., (2007). Evaluating the aerobic biodegradability of plastics in soil environments through GC and IR analysis of gaseous phase. *Polym. Test., 26*, 729–741.
63. Berthe, L., Druilhe, C., Massiani, C., Tremier, A., & De Guardia, A., (2007). Coupling a respirometer and a pycnometer, to study the biodegradability of solid organic wastes during composting. *Biosyst. Eng., 97*, 75–88.
64. Grima, S., Bellon, M. V., Feuilloley, P., & Silvestre, F., (2000). Aerobic biodegradation of polymers in solid – state conditions: A review of environmental and physicochemical parameter settings in laboratory simulations. *J. Polym. Environ., 8*, 183–195.
65. De Guardia, A., Petiot, C., & Rogeau, D., (2008). Influence of aeration rate and biodegradability fractionation on composting kinetics. *Waste Manag., 28*, 73–84.
66. Sundberg, C., & Jönsson, H., (2008). Higher pH and faster decomposition in biowaste composting by increased aeration. *Waste Manag., 28*, 518–526.
67. Száraz, L., & Beczner, J., (2003). Optimization processes of a CO 2 measurement set – up for assessing biodegradability of polymers. *Int. Biodeterior. Biodegradation, 52*, 93–95.
68. Liang, C., Das, K. C., & McClendon, R. W., (2003). The influence of temperature and moisture contents regimes on the aerobic microbial activity of a biosolids composting blend. *Bioresour. Technol., 86*, 131–137.
69. Pagga, U., Beimborn, D. B., Boelens, J., & De Wilde, B., (1995). Determination of the aerobic biodegradability of polymeric material in a laboratory controlled composting test. *Chemosphere, 31*, 4475–4487.
70. Bellia, G., Tosin, M., Floridi, G., & Degli, I. F., (1999). Activated vermiculite, a solid bed for testing biodegradability under composting conditions. *Polym. Degrad. Stab., 66*, 65–79.

71. Chiellini, E., & Corti, A., (2003). A simple method suitable to test the ultimate biodegradability of environmentally degradable polymers. *Macromol. Symp., 197,* 381–396.

72. Degli, I. F., Tosin, M., & Bastioli, C., (1998). Evaluation of the biodegradation of starch and cellulose under controlled composting conditions. *J. Polym. Environ., 06,* 197–202.

73. Solaro, R., Corti, A., & Chiellini, E., (1998). A new respirometric test simulating soil burial conditions for the evaluation of polymer biodegradation. *J. Polym. Environ., 06,* 203–208.

74. Starnecker, A., & Menner, M., (1996). Assessment of biodegradability of plastics under simulated composting conditions in a laboratory test system. *Int. Biodeterior. Biodegradation, 37,* 85–92.

75. Tosin, M., Degli, I. F., & Bastioli, C., (1998). Detection of a toxic product released by a polyurethane – containing fi lm using a composting test method based on a mineral bed. *J. Polym. Environ., 06,* 79–90.

76. Way, C., Wu, D. Y., Dean, K., & Palombo, E., (2010). Design considerations for high – temperature respirometric. *Polym. Test., 29,* 147–157.

77. Standards, (2005). AS ISO 14855–2005. Plastic *Materials – Determination of the Ultimate Aerobic Biodegradability and Disintegration under Controlled Composting Conditions – Method by Analysis of Evolved Carbon Dioxide.* Standards Australia, Sydney, NSW Australia.

78. Rhim, J. W., Hong, S. I., Park, H. M., & Ng, P. K. W., (2006). Preparation and characterization of chitosan – based nanocomposite films with antimicrobial activity. *J. Agric. Food Chem., 54,* 5814–5822.

79. Fukushima, K., Abbate, C., Tabuani, D., Gennari, M., & Camino, G., (2009). Biodegradation of poly(lactic acid) and its nanocomposites. *Polym. Degrad. Stab., 94,* 1646–1655.

80. Shimpi, N., Borane, M., Mishra, S., & Kadam, M., (2012). Biodegradation of polystyrene (PS) – poly(lactic acid) (PLA) nanocomposites using *Pseudomonas aeruginosa. Macromol. Res., 20,* 181–187.

81. Shih, Y. F., & Wu, T. M., (2009). Enzymatic degradation kinetics of poly(butylene succinate) nanocomposites. *J. Polym. Res., 16,* 109–115.

82. Arena, M., Abbate, C., Fukushima, K., & Gennari, M., (2011). Degradation of poly(lactic acid) and nanocomposites by Bacillus licheniformis. *Environ. Sci. Pollut. Res., 18,* 865–870.

CHAPTER 11

Mechanisms of Action of Multifunctional Media Submitted by Nano- and Ultrafine Iodine Compounds on Pathogens of Bacterial and Fungium Etiology

V. B. GOLUBCHIKOV,[1] A. V. ZHIVOTKOV,[1] and A. V. VAKHRUSHEV[2,3]

[1]Scientific and Production Company "Nord," Perm, Russia,
E-mail: nord59r@mail.ru

[2]Department of Mechanics of Nanostructures, Institute of Mechanics, Udmurt Federal Research Center, Ural Division, Russian Academy of Sciences, Izhevsk, Russia

[3]Department of Nanotechnology and Microsystems, Technic Kalashnikov Izhevsk State Technical University, Izhevsk, Russia

ABSTRACT

The chapter describes the results of studies on the effects of special multifunctional media based on nano-ultrafine aerosols of iodine compounds on pathogens of bacterial etiology. Research conducted in the laboratory, hospital, production of poultry, and animals. In the process of research, the previous experience of scientists was considered, international techniques were applied. The research results showed the high efficiency of nano-ultrafine media on pathogens.

11.1 THEORETICAL JUSTIFICATION

A number of disinfectants, for example, oxidizers (chlorine, iodine, and their compounds, hydrogen peroxide, potassium manganese acid), mineral salts (sulfurous, boric, fluoride-hydrogen), because active oxidative processes that are not characteristic of the metabolism of pathogen cells. Additionally, these substances block the work of the enzyme systems of pathogens, which together leads to cell death.

Many chemicals are used in medicine, agriculture, food industry, as disinfectants, they are:

- Cause rapid (within a few minutes) cell death of pathogens (disinfectants are more active in environments poor in organic matter);
- Destroy not only vegetative cells, but also spores;
- Do not cause the emergence of resistant forms of microorganisms.

In the food industry, substances containing active chlorine (chloramine, bleach) are used as disinfectants. In medicine, iodine compounds and hydrogen peroxide are widely applicable. Formaldehydes are used in agriculture.

Since the beginning of the last century, domestic, and foreign scientists have paid attention to iodine and its compounds as a multifunctional agent capable of fighting pathogens of various etiologies without causing the effect of resistance [1–5].

In their work, the team of specialists of NPF NORD LLC and the Udmurt Federal Research Center set the task of creating and implementing a pathogen control tool that is distinguished by a number of features:

- The safety of the use of funds. The possibility of varying concentrations and use in habitable conditions (people, animals, plants) and uninhabited objects. Absence: residual toxic effect, the possibility of food processing and their further sale;
- The introduction of the controllable gaseous media CGM by the volumetric method in the form of a dry aerosol of condensed particles. Creating a uniform concentration over the entire volume of the object, including shaded areas, vertical, and horizontal niches;
- Low labor costs provided the method of making a controlled gas environment.

The result of the development work and a wide range of literature and experimental research was the creation of products:

- Mineral complex "x";
- Generator of multifunctional media "iodine."

Which were the implementation of the principle of several authors, including [6], where thermo-projectile drafts are described.

The products are based on the use of a mixture of a heat-generating composition compressed into a cylindrical block, into which iodine compounds are introduced. When you activate the checkers, by exposure to the flame, the synthesis begins:

- **Transport Substances:** Carbon dioxide (CO_2) and nitrogen (N_2), whose task is the uniform distribution of the active substance by volume;
- **Active Substance:** Iodine compounds acting on pathogen cells of various etiologies.

The whole mixture is an aerosol with a fraction of nano-and ultra-dispersed particles – nano-/ultrafine aerosol (Figure 11.1).

FIGURE 11.1 Generators multifunctional media "iodine."

In Figure 11.1, a photo of generators of various sizes: 22 grams, 85 grams and 600 grams of the active substance, which allows you to process different amounts of volume, create different concentrations for the purpose of: air rehabilitation, prevention (control of the pathogen population) and disinfection (fighting pathogens) is presented.

Figure 11.2 presents a constructive scheme of the mineral complex "X." In the process of the heat-generating composition, the sublimation of the active substance takes place. The decomposition products of the heat-generating composition and the active substance are mixed, and the resulting medium is distributed throughout the object being processed (Figure 11.3).

heat-generating composition   active substance

FIGURE 11.2   Mineral complex "x."

FIGURE 11.3   The process of synthesis of CGS. Mineral complex "X."

The process of synthesis of the active substance and transport substances is presented in Figure 11.3. The dark color in the aerosol is the active substance transferred to the aerosol; the light component is the decomposition products of the heat-generating composition. The gray cylinder on the right is the mineral complex "X" upon completion of the synthesis process.

11.2 RESULTS OF EXPERIMENTAL RESEARCHES

From here on, under the CGS will be understood the controlled atmosphere, consisting of the transport component – CO_2 and N_2 gas, as well as the active substance – iodine compounds.

The purpose of the research was to determine the effect of CGS on the objects of study – pathogens of the following etiologies (in laboratory and production conditions):

- **Bacteria:** Marker pathogens;
- **Bacteria:** Pathogens.

11.2.1 ACTION OF NANO-ULTRAFINE AEROSOL ON PATHOGENS OF BACTERIAL ETIOLOGY

In the laboratory, research was conducted on the effects of CGS aerosol on two marker pathogens:

- *Staphylococcus* 906, which characterizes the resistance of the intestinal group of bacteria;
- *Escherichia coli* 1257, the most resistant species from the cocci group of microbes;

As on the pathogen *Pseudomonas aeruginosa* – a pathogen that causes sepsis resistant to antibiotics.

In this section and further, the materials and methods described in Refs. [7–12] are used.

The conditions of the experiment:

1. For each pathogen, optimal growth conditions are provided, including temperature control and nutrient regimes.
2. The microbial load for each of the crops sown is 10^7 (pcs.)/sm^2.

3. For the experiment, control (without CGS) and experimental (with CGS) groups are used. Environmental conditions for temperature and humidity for the experimental and control groups are similar.
4. Exposure time 30 minutes.
5. The result was evaluated 24–72 hours after exposure.

The research results are shown in Table 11.1 and in the histograms of Figure 11.4 (a, b, c).

TABLE 11.1   Pathogen Microbial Load

Pathogen	Concentration "X"		Concentration "2X"	
	Control	**Experimental**	**Control**	**Experimental**
Staphylococcus 906	$10^7 \times sm^{-2}$	$10^2 \times sm^{-2}$	$10^7 \times sm^{-2}$	Absent
Escherichia coli 1257	$10^7 \times sm^{-2}$	$10^4 \times sm^{-2}$	$10^7 \times sm^{-2}$	$10^2 \times sm^{-2}$
Pseudomonas aeruginosa	$10^7 \times sm^{-2}$	$10^6 \times sm^{-2}$	$10^7 \times sm^{-2}$	$10^2 \times sm^{-2}$

The graphic materials below show the results of colony growth under laboratory conditions on samples of the experimental (processed) and control (without treatment) groups. Samples were seeded in Petri dishes, each culture for the optimal nutrient solution.

In a series of experiments for the concentration of "X" used cups with a wall, where in one half, for samples of the experimental and control groups, were sown, and the other not. This is done to confirm the level of sterility in the laboratory due to the absence of any colonies in the non-sown area upon completion of the incubation.

It should be noted that for marker pathogens: *Staphylococcus*, *Escherichia coli*, the use of nano-ultrafine aerosols based on iodine compounds resulted in a significant decrease in colony growth. Bacterial infection is reduced from 10^3 (thousands) to 10^5 (one hundred thousand) times. In one experiment, the growth of *Staphylococcus* colonies was stopped completely.

A high result was achieved when exposed to *Pseudomonas aeruginosa*, the pathogen characterized by high resistance to antibiotics. If in a series of experiments with a concentration of "x" infection was reduced by a factor of 10, then as the concentration increased to "2x," the number of colonies decreased by 10^5 (one hundred thousand) times.

Consider in more detail the figures. In Figure 11.5, for the sample "experimental" and "control," the right half of the surface is sown, to the

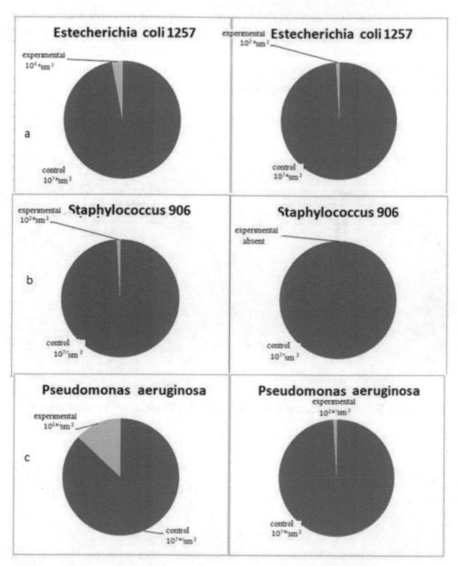

FIGURE 11.4 a) *Staphylococcus* 906. The ratio of microbial contamination for experimental and control samples with exposure concentrations of "x" (left) and "2x" (right). b) *Escherichia coli,,* 1257. The ratio of microbial contamination for experimental and control samples with exposure to concentrations of "x" (left) and "2x" (right). c) *Pseudomonas aeruginosa.* The ratio of microbial contamination for experimental and control samples with exposure to concentrations of "x" (blue) and "2x" (orange).

right of the partition line. On the "control" sample, a uniform matt surface covered with a cultured colony is clearly visible, while the part to the left of the partition is evenly glossy. For the "experimental" sample, the seeded surface is glossy with local patches of sprouted colonies. This is more expressed for the pathogen *Staphylococcus* (Figure 11.6).

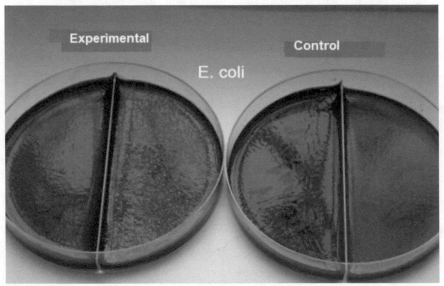

FIGURE 11.5 Experienced and control sample of *Escherichia coli*. Concentration "X."

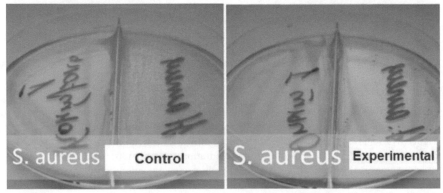

FIGURE 11.6 Experienced and control sample of sowing *Staphylococcus*. Concentration "X."

When considering the results in Figure 11.7 for the blue pus bacillus is visible: the "control" sample uniformly covers the dark layer of the cultured colony. It resembles a dense film of bitumen. The sample "experimental" colonies further expanded that the rest of the surface was not suitable for their development.

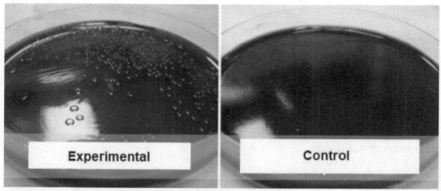

FIGURE 11.7 Experimental and control sample of seeding *Pseudomonas aeruginosa.* Concentration "2X."

Most likely, the growth of colonies was possible in those places where the nano-ultrafine particles of the active substance were not delivered by the gaseous medium.

After carrying out the laboratory cycle of research (the results in Table 11.1 and in Figures 11.4–11.7), tests in the autopsy department of the Institute of Pathology and Bacteriology of the Kaiser Franz Josef Hospital, Vienna was performed.

The following cultures were present as pathogens: *Staphylococcus aureus, Enterococcus fecallis, Escherichia coli, Klebsiella pneumonia, Pseudomonas aeruginosa, Clostridum diffcile,* as well as the fungal strain *Candida albicans,* sown on the following surfaces: surfaces of working hospital gowns, plastic covering of tables.

1. The initial concentration of pathogens on surfaces was$> 10^7$ * sm^{-2}.
2. The exposure time was 2.5 hours.
3. The concentration of active substances – "2x."
4. The exposure time of pathogens after exposure (cultivation time)– 48 hours.

At the time of evaluation of the results of the experiments, washings were taken from the surfaces under consideration. The data obtained showed that after exposure, microbial contamination decreased 10,000 times and amounted to $<10^3 \times sm^{-2}$, which confirmed the correctness of the choice of concentration in the laboratory.

A common effect for all pathogens is the presence of growth-free zones where the colonies have not grown and the seeding on which does not lead to the growth of new colonies. Processing surfaces with nutrient media stops the growth of colonies, and not just temporarily slows them down.

11.2.2 RESULTS OF MICROBIOLOGICAL TESTS

Bacteria of *Escherichia coli* and *Staphylococcus aureus* were considered as a marker pathogen.

1. The opening room was considered as the processed room.
2. Controlled surfaces: table, floor, door, wall (tile), wall (paintwork).
3. The concentration of the active substance "x."

In addition, air sampling was carried out (Table 11.2).

TABLE 11.2   Results of Microbiological Tests

Surface	*Escherichia coli*		*Staphylococcus aureus*	
	Before disinfection	**After**	**Before disinfection**	**After**
1. Floor	*Escherichia coli* $10^4 \times sm^{-2}$	Absent	*Escherichia coli* $10^4 \times sm^{-2}$	Absent
2. Door	*Escherichia coli* $10^2 \times sm^{-2}$	Absent	*Escherichia coli* $10^2 \times sm^{-2}$	Absent
3. Wall (tile)	*Escherichia coli* $10^2 \times sm^{-2}$	Absent	*Escherichia coli* $10^2 \times sm^{-2}$	Absent
4. Wall (paintwork)	*Escherichia coli* $10^2 \times sm^{-2}$	Absent	*Escherichia coli* $10^2 \times sm^{-2}$	Absent

With initially low microbial contamination, not more than $10^4 \times sm^{-2}$, complete destruction of bacterial pathogens with small concentrations of the active substance was performed, which may indicate the possibility of periodic preventive treatment of the premises, which will not allow the cultivation of pathogens to high concentrations (10^7 and above).

11.2.3 MICROBIOLOGICAL RESEARCH

The use of CGS nano-ultrafine aerosols for the rehabilitation of poultry houses. When conducting practical tests of the bactericidal properties of CGS based on iodine compounds during the rehabilitation of the poultry house in the presence of broiler chickens, it was established that after treatment in washes taken from the surface of walls, feeding troughs and other technological equipment, no bacteria of the genus *Staphylococcus* were detected (80% of the total swabs) and *E. coli* (100% of the total number of swabs).

Disinfection in the house with a volume of 9500 m³ was carried out in the presence of 24 thousand heads of broiler chickens 37 days old. Sources CSG were located evenly in ten points of the room. During the synthesis, an aerosol was formed, which evenly filled the entire premises of the house. The exposition of the CGS in the house was 30 minutes (Table 11.3).

TABLE 11.3   The Effectiveness of the Sanitizing Action of the CGS in the Disinfection of the House for Growing Broiler Chickens

Parameter	Before disinfection	After disinfection
General microbial contamination of air, KOE/м³	191,323	141,138
E. coli contamination in the air, KOE/м³	6057	3454

According to the results of the sanitizing action of the CGS aerosol on pathogens in the air, the total microbial contamination decreased by 30%, the content of *Escherichia coli* > 40% (Figures 11.8 and 11.9).

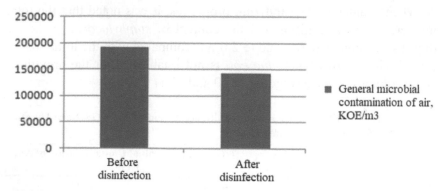

FIGURE 11.8   General microbial contamination of air.

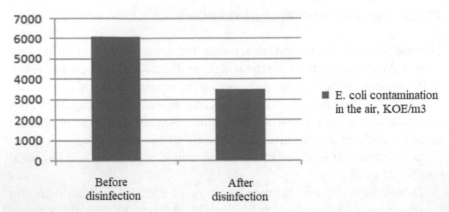

FIGURE 11.9   *E. coli* contamination in the air.

Also, in the process of disinfection, there were no changes in the clinical condition of the bird (anxiety, coughing, sneezing, and other pathological reactions).

Experimental studies on the effects on pathogens in pigsty conditions in the presence of animals have been carried out.

The drug was used for prophylactic dry disinfection of the room. Disinfection was carried out in two sectors of the area for rearing piglets in the presence of 529 and 537 heads of pigs 64 and 68 days old. The drug was placed evenly at two points in each sector. The exposure of the CGS aerosol is 30 minutes (Table 11.4).

The quality control of disinfection was carried out by the presence on the surfaces of the treated premises of viable cells of sanitary indicative microorganisms (*Escherichia coli* and *staphylococci*).

When evaluating the sanitizing properties, it was noted that the total number of microorganisms and the content of *staphylococci* in the air after disinfection were reduced to 2 times compared with the initial background. Growth of *Escherichia coli* is not established, or the growth of single colonies is noted (Figures 11.10 and 11.11).

TABLE 11.4   The Effectiveness of the Sanitizing Action of the CGS Nano-Ultrafine Aerosol Disinfection Pigsty for Growing Pigs

Active Substance Concentration	General Microbial Contamination of Air, KOE/м³	
"X"	44,762	22,698
"2X"	296,345	219,177

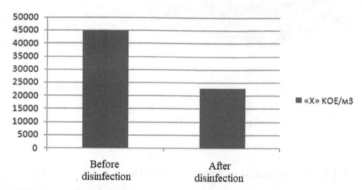

FIGURE 11.10 General microbial contamination of air KOE/м³. Active substance concentration "X."

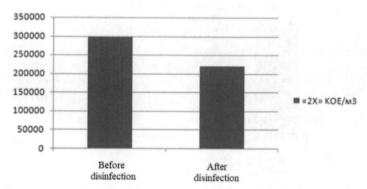

FIGURE 11.11 General microbial contamination of air KOE/м³. Active substance concentration "2X."

Below (Table 11.5, Figures 11.12 and 11.13) results of researches on calf disinfection are presented. The concentration (active substance) of the aerosol "X." Processing in the presence of animals.

Disinfection was carried out in the presence of 400 heads of calves aged from one month to a year.

TABLE 11.5 The Effectiveness of the Sanitizing Action of CGS During Disinfection of Calves

Parameter	Before Disinfection	After Disinfection
General microbial contamination of air, KOE/м³	72,500	30,750
E. coli contamination in the air, KOE/м³	27,500	2,800

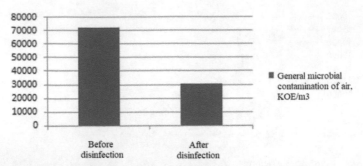

FIGURE 11.12   General microbial contamination of air in the calf house.

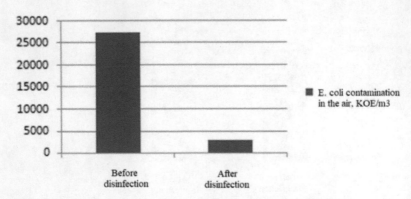

FIGURE 11.13   *E. coli* microbial contamination of air in the calf house.

In the process of operation of the generator of multifunctional media CGS nano-ultrafine aerosol evenly filled calf house. The exposition was 30 minutes. Quality control, as in the previous experiments, was carried out by the presence of viable marker pathogens on the surfaces of the treated premises. An assessment was made of the total microbial contamination and the content of *E. coli* in the air.

It was found that after exposure, no bacteria of the genus *Staphylococcus* and *E. coli* were detected in washings taken from the surface of the enclosing structures (floor, walls, and feeding trough). When evaluating the sanitizing properties, it was noted that the total number of microorganisms in the air after disinfection was reduced by two times compared with the initial bacterial background. In addition, there was a significant decrease in *E. coli* in the air (10 times) compared with the initial level. In

50% of the collected air samples of the growth of *Escherichia coli*, either no single colonies were observed, or growth was observed.

Also, in the process of disinfection is a not marked change in the clinical condition of the animals (anxiety, coughing, sneezing).

11.3 CONCLUSION

1. Completed the study to determine the bactericidal properties of CGM regarding pathogens of bacterial etiology in the laboratory and in the conditions of production.
2. High efficiency of CGM was noted in destroying and stopping the growth of colonies of marker pathogens of the genus *Staphylococcus* and *Escherichia coli*, where contamination decreased 10^5 (one hundred thousand times) in individual experiments or colonies were completely destroyed.
3. The number of pathogens in the air, after treatment, decreased by two times or more, which favorably affected the conditions of animals.
4. The above treatments were carried out in the presence of animals (birds, piglets, calves). As a result of the treatment, a decrease in the mortality of animals of childhood was observed by 2–4 times. Changes in the clinical state of the animals during and after treatment were not observed.
5. Separately, the effect of CGM on *Pseudomonas aeruginosa*, an antibiotic-resistant pathogen, was studied. As a result of the treatment, it was possible to reduce the contamination of the surface 10^5 (one hundred thousand) times and completely stop the growth of colonies.
6. The use of CGM based on iodine compounds and the method of treatment showed high efficiency coupled with low labor costs.
7. Additionally, the use of CGM showed the absence of a negative, including the corrosive and cosmetic effect on the finishing elements, enclosing structures, equipment.

The task of further research is to expand the application of the described technology in various fields of science and technology-based on theoretical modeling [17] and applied research [18] of the authors.

ACKNOWLEDGMENTS

The work was carried out with the support of NORD Research and Production Company LLC, Ural Branch of the Russian Academy of Sciences and Technic Kalashnikov Izhevsk State Technical University.

KEYWORDS

- controlled atmosphere
- disinfection
- nano-ultrafine aerosol
- pathogens
- thermo-projectile drafts

REFERENCES

1. Burnet, F. M., Holden, H. F., & Stone, J. D., (1945). Action of iodine vapors on influenza virus in droplet suspension. *Austral, J. Sci., 7*, 125–126.
2. Crosby, A. W., (1918). *America's Forgotten Pandemic: The Influenza of 1918* (pp. 132–148). New York, Cambridge University.
3. Carroll, B., (1955). The relative germicidal activity of triiodide and diatomic iodine. *J. Bacteriol., 69*, 413–417.
4. Gottardi, W., (1991). Iodine and iodine compounds. In: Block, S. S., (ed.), *Disinfection, Sterilization, and Preservation* (Vol 2, pp. 447–454). Philadelphia, Lea & Febiger.
5. Derry, D., (2009). *Iodine: The Forgotten Weapon Against Influenza Viruses Thyroid Science, 4*(9), 1–5.
6. Solodnikov, S. Y., (2006). In: Solodnikov, S. Y., & Solov, I. V., (eds.), *Thermo-Projectile Drafts* (Vol. 5, pp. 15–18). Veterinary.
7. Methodical instructions for quality control of disinfection and sanitization of facilities subject to veterinary and sanitary inspection, approved by the Ministry of Agriculture of the Russian Federation, (2002). (15.07.2002 N 13–5–2/0525). p. 65.
8. MUK No. 859–70 "Guidelines for assessing the effectiveness of disinfectants intended for disinfecting various objects and sanitizing people," (1970). p. 21.
9. Instruction No. 737–68 "By definition, the bactericidal properties of new disinfectants," approved by the Ministry of Health of the USSR, (1968). p. 16.
10. Methods described in R4.2.2643–10 "Methods of testing disinfectants for assessing safety and efficacy," (2010). p. 611.
11. MUK 4.2.2316–08 "Methods for the control of bacteriological nutrient media. Methodical instructions, (2008). p. 67.

12. Instructions on the procedure and frequency of monitoring the content of microbiological and chemical pollutants in meat, poultry, eggs and their products, (2000). p. 76.

13. Instructions for sanitary and microbiological control of carcasses, poultry meat, poultry products, eggs and egg products at poultry and poultry processing plants, (1990). p. 45.

14. Anufriev A. N., (2004). Safe production of aerosols of disinfectants and drugs. In: Anufriev, A. N., et al. (eds.), *Veterinary Medicine* (Vol. 8, pp. 7, 8).

15. Bochenin, Y. I., (2004). Aerosols in the prevention of infectious diseases of farm animals. In: Yu, I. B., et al., (eds.), *Veterinary Consultant* (Vol. 23, 24, pp. 10–18).

16. Birman, B. Y., (2007). Guidelines for aerosol disinfection of poultry premises. In: Ya, B. B., Vysheleskogo, S. N., et al., (eds.), *Minsk, RNIIUP "IEV Them* (p. 56).

17. Veterinary and sanitary rules for veterinary disinfection. Guidelines for quality control of disinfection and sanitization of objects subject to veterinary supervision: Sat. regulatory documents on veterinary medicine./Ch. ex. Veterinary with GOS, Veterinarian and GOST, (2007). In: Ansenov A. M., et al., (eds.), *Prod. Ins. Minsk* (p. 96).

18. Vakhrushev, A. V., (2017). *Computational Multiscale Modeling of Multiphase Nanosystems. Theory and Applications.* Apple Academic Press: Waretown, New Jersey, USA.

19. Alikin V. N., Vakhrushev A. V., Golubchikov V. B., Lipanov A. M., & Serebrennikov, S. Y., (2010). *Development and Investigation of the Aerosol Nanotechnology* (p. 196). Moscow: Mashinostroenie (in Russian).

CHAPTER 12

Methods of Obtaining and Application of Graphene

A. V. GUMOVSKII[2] and A. V. VAKHRUSHEV[1,2]

[1]Department of Mechanics of Nanostructures, Institute of Mechanics, Udmurt Federal Research Center, Ural Division, Russian Academy of Sciences, Izhevsk, Russia

[2]Department of Nanotechnology and Microsystems, Technic Kalashnikov Izhevsk State Technical University, Izhevsk, Russia, E-mail: gumma.andres@gmail.com

ABSTRACT

The chapter describes the main methods for producing graphene (GR), such as the CVD method (chemical gas-phase deposition), epitaxial growth on silicon carbide or metals, mechanical exfoliation, etc. It is noted that the CVD method is the most promising, relatively inexpensive, and affordable method for producing GR high enough quality. A schematic representation of the setup for CVD synthesis is given, the CVD synthesis process is described. Several areas of application of GR are shown: biofuel elements (BFC) and GR -based electrode materials, GR as a sensor membrane, GR as a superconductor in electronics, the use of GR in capacitors.

12.1 INTRODUCTION

The ability of carbon to form complex chains is fundamental to organic chemistry and the basis for all life on Earth, which makes it a unique element of the Periodic Table. Until 2004, three-dimensional (3D, diamond, graphite), one-dimensional (1D, nanotubes) and zero-dimensional (0D, fullerenes) allotropic forms of carbon were known. For a long time, two-dimensional forms of carbon (or 2D-graphite) could not be obtained experimentally, according to the arguments of Landau and Peierls that strictly 2D-crystals are thermodynamically unstable. Earlier attempts were made to grow graphene (GR) or isolate it using the chemical peeling method, and only in 2004, using advanced micromechanical spallation techniques, was it possible to get together with physicists from the University of Manchester (Britain) under the leadership of Andre Geim and Konstantin Novoselov using conventional tape – Scotch tape for sequential separation of layers from ordinary crystalline graphite. Scientists based on the fact that GR becomes visible in an optical microscope if it is placed on the surface of a silicon substrate with a certain thickness of the SiO_2 layer, and this simple but effective way to scan the substrate in search of GR crystals was the determining factor for their success.

Obviously, GR is one of the most interesting modifications of carbon. This is the thinnest material: the structure of GR is represented by a crystal lattice, one carbon thick.

In addition, GR is one of the most durable materials and its resistance to mechanical stress is comparable to that of diamond, but it bends well and easily folds into a tube, which makes it an ideal material for the manufacture of nanotubes – structures that are used to simulate various natural processes.

All these amazing properties of GR arise from the uniqueness of its charge carriers, which behave like relativistic particles.

Another effect due to the nature of charge carriers in GR is associated with the presence of helicity, which leads to the existence of so-called chiral symmetry (translated from the Greek word "cheir" – hand). The chiral nature of electron states in single-layer, and two-layer GR plays an important role in the passage of an electron through a potential barrier (tunnel effect). Now, there are several basic methods for obtaining GR, in addition to physical exfoliation.

12.2   METHODS FOR PRODUCING OF GRAPHENE (GR)

GR is a flat sheet, one atom thick, and carbon atoms with SP_2 hybridization, located in a hexagonal lattice. It is believed that this will be "the thinnest and most durable material in the universe" and predicted that it would have remarkable physical and chemical properties. Nevertheless, the internal properties of GR depend on its structural perfection, which strongly depends on the methods of synthesis. GR can be obtained in many ways, such as micromechanical exfoliation of graphite [1], epitaxial growth on silicon carbide or metals [2], chemical deposition on metal substrates [3, 4], thermal or chemical reduction of graphene oxide (GO) manufactured by oxide separation graphite [5], and the detachment of intercalated graphite compounds (GIC) [6, 8]. Among these methods, micromechanical cleavage is currently the most efficient and reliable method for producing high-quality GR.

Epitaxial growth graphene (EPGG) and chemical vapor deposition graphene (CVDG) also have high quality and excellent physical properties [2, 4, 7]. However, with their help, it is difficult to obtain a large amount of GR in order to satisfy the need for quality composite fillers. Since exfoliation of GIC is a technique of physical detachment, graphene obtained by exfoliating intercalated graphite compounds (GICG) is quite possible to retain the structure of GR and be produced in large quantities [6]. However, in most cases, the GICG obtained consists of multilayer sheets due to the conversion of GR layers after deintercalation and the presence of numbered phases of a high GIC stage. Intercalation is also a problem [6, 9].

The most promising, relatively inexpensive and affordable method for producing GR of sufficiently high quality is chemical gas-phase deposition (CVD) on the surface of transition metals such as Ni, Pd, Ru, Ir, Cu, etc. This method was studied and used even before the discovery of GR. The formation of GR structures (thin graphite) as a result of the preparation of transition metal surfaces and in industrial heterogeneous catalysis has been known for almost 50 years [10].

The synthesis of GR was carried out in an installation schematically depicted in Figure 12.1. A thermal reactor consists of a furnace (thermally

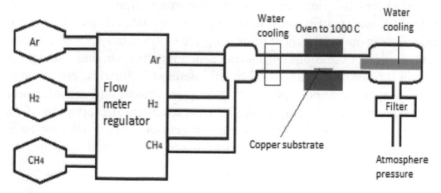

FIGURE 12.1 Schematic representation of the setup for CVD synthesis.

insulated casing with a heater). It contains a quartz cylindrical tube. The tube can be moved in the up and down direction of the gas mixture. This is necessary to remove the catalyst substrate from the hot zone (synthesis zone) in order to carry out a sharp cooling of the substrate.

Graphitization of the surface of metals was used to change physical properties and prevent corrosion. Layers of graphite were first found on Ni surfaces [10, 11, 13], which were exposed to carbon sources in the form of hydrocarbons or gaseous carbon. Nowadays, polycrystalline GR films of large sizes are obtained by CVD. The advantage of this method is the scalability of the obtained samples. The difficulties of this method are associated with controlling the growth of a single layer and the presence of defects in the resulting material. Also, the disadvantage of this method is the need to transfer the GR film grown on the metal surface to the desired surface. In the process of film transfer, such methods as vacuum, chemical, and electrochemical etching of metal substrates are used [12].

The growth mechanism of the film is associated with two processes. The first is the thermal decomposition of carbon-containing gases on the surface of metals. The second is the dissolution of carbon in the metal at high temperatures and the subsequent segregation (release) on the surface during cooling. The solubility of carbon in the metal, the crystal lattice of the surface and the conditions of the growth process determine the morphology and thickness (number of layers) of the GR film. Growth on a hexagonal lattice is often called epitaxial, even if there is no significant agreement between the lattice and the substrate [12].

In order to take full advantage of its properties for various applications, except for electronic devices, the integration of individual GR into polymer matrices to form modern multifunctional composite materials is one of the most promising areas, since polymer composites usually have exceptional properties, specific strength, and widespread use in the aerospace, automotive, and defense industries, etc. In addition, polymer composites can be easily processed and manufactured into components of complex shape with excellent preservation of the structure and properties of GR using conventional processing methods. It is very important to take full advantage of the outstanding properties of GR.

12.3 APPLICATION OF GRAPHENE (GR)

12.3.1 BIOFUEL ELEMENTS (BFC) AND GRAPHENE (GR) NANOMATERIAL FOR ELECTRODES

The first successes of using GR were obtained when designing biosensors [15]; the data showed that GR could be effectively used for conjugation with the biomaterial and in fact opened the way to its use in biofuel elements (BFC). Nevertheless, it is rather surprising that by now only a few examples of the use of GR as part of enzyme BFC have been described. The sharing of microbial cells and GR is being studied more intensively; apparently, this situation is due to the fact that, in general, the total number of publications on microbial BFC is three times higher than on enzyme BFC. In 2009, the first works on the possibility of the conjugation of glucose oxidase (GOx) and GR were published [16]; a year later, publications appeared on the use of GOx and GR as part of BFC [17].

Laccase was mainly used as a cathode catalyst. GOx contains flavin adenine nucleotide, a cofactor surrounded by a protein and a glycan structure that limits the effective exchange of electrons between the protein's active center and the surface of the electrode; thus, the structure of the enzyme creates a barrier to the functioning conditions of the bioelectrode. It should also be noted that even if a non-mediator transfer of electrons for a biosensor containing a GOx is described, this does not mean that the catalytic current is necessarily associated with a direct transfer between the cofactor and the electrode. This effect may be the result of a non-enzymatic reaction of hydrogen peroxide or oxygen participating in the catalytic cycle of the GOx on the electrode surface [18].

On the other hand, direct transfer is known for the cathode application of the GOx, when a current is generated during the reduction process of the FAD-cofactor, which is then re-oxidized with oxygen. Obviously, this case is not acceptable for the operation of BFC [19]. This situation, of course, allows the design of year-based biosensors, but the principle is not applicable to the creation of BFC for the following reasons:

1. GOx more effectively transfers electrons to oxygen than to electrodes, which can be judged by the value of the electron transfer constant.

2. The resulting hydrogen peroxide is not an acceptable substrate. This reaction, competitive to oxygen, is the most important in essence. It should be taken into account when there is a question about the use of the GOx in BFC.

Even in the case of direct or, more precisely, without the mediocre transfer of electrons between the GOx and the electrode, the mechanism cannot be proved with absolute accuracy, and therefore there is always a counter-version of the reality of direct transfer of electrons. In this regard, it should be assumed that direct anodic electronic exchange between the GOx and the electrode is prohibited, and only the transfer is allowed (see Figure 12.2). This reasoning is confirmed by the following example of BFC: the anode consisted of the GOx enzyme immobilized on GR oxide and a glassy carbon electrode, the cathode was fixed by laccase on multi-layer carbon nanotubes modified by zinc oxide. BFC had a no-load voltage of 60 mV, a maximum power density of 0.054 $\mu W*cm^{-2}$ and a voltage at a maximum power of 50 mV [14].

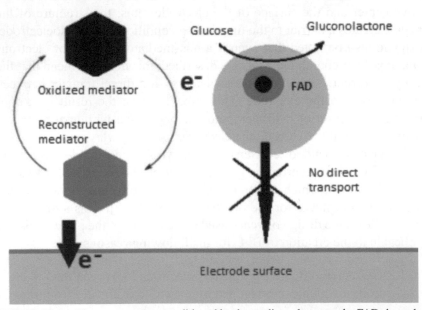

FIGURE 12.2 Electron transport conditioned by the mediator, between the FAD-dependent GOx and the electrode for the case when the direct transfer is not possible. The FAD cofactor is indicated by gray and black circles.

At the same time, the example of BTE (anode configuration-GR/Carbon Nanotube-COOH/GOx, cathode configuration-GR/Carbon Nanotube-COOH/laccase + ABTS (2,2'-azino-bis- (3-ethylbenzthiozolin-6-sulfonic acid)) in solution, the no-load voltage is 1200 mV, the maximum specific power is 2270 $\mu W \ cm^2$, the voltage at the maximum power is 500 mV [20]) contradicts the described scheme of the mechanism of functioning of the GOx, since the parameters of this BFC are the best-studied for enzyme BFC [16].

12.3.2 GRAPHENE (GR)-BASED MICROBIAL BIOFUEL CELLS

Although the idea of producing electricity from the oxidation of organic substrates by the microbial biocatalyst was first formulated more than a hundred years ago [21], it took decades for important results to be obtained in the field of microbial BFC. The task of ensuring efficient electron transport between the electrode surface and the enzyme localized inside the microbial cell turned out to be quite complex. The challenge was not only to ensure efficient charge transport, but also to ensure the transport of the substrate to the biocatalyst; for the microbial cell, it was much harder than for the enzyme. As a result, the maximum powers of BFC, based on enzymes, were higher than those of microbial BFCs. In this regard, in the early 1980s, the main focus of the research was the use of electronic carriers – mediators [68]. However, two decades later, there were reports of the possibility of mediate transfer [22, 23], and since that period, studies of microbial BFC have acquired a new scope [24]. It was established that bacterial cells could have three main electron exchange pathways with electrodes — using secreted mediators, using cytochromes, and using bacterial pili or nanowires (Figure 12.3) [25–27]. However, even with the above assumptions regarding the exchange mechanisms, further study of the details was required. Thus, the question arose whether the transfer occurs by a mechanism of transfer from one cytochrome site to another or is transferred by the conduction mechanism in metals through bound electrons in the aromatic rings of amino acids [25]. It should also be noted that the combination of the three transfer mechanisms for bacterial cells introduces its difficulties in the unambiguous interpretation of the process of electron transfer [28].

Regardless of the existing complexity of explaining the transfer mechanisms, the discovery of the effect of direct electron transport on bacterial cells provided ample opportunities for constructing BFC. Such bacteria are called "exoelectrogenic," i.e., capable of independently generating electric energy and containing nanosystems, which can be used to create non-reagent microbial BFC. The power of BFC on their basis reaches the power of many BFC based on enzymes. The use of GR materials in microbial BFC does not allow achieving the best power values of BFC based on enzymes. At the same time, microbial BFC has other positive qualities; for example, a significantly higher operating stability, an unusually wide range of substrates [29]. Currently, much attention is paid to the use of conductive nanomaterials (NMs) (carbon nanotubes, carbon, and polymer nanofibers, graphite particles) and conductive macro-dimensional materials – carbon cloth, carbon paper, carbon felt, including on the basis of exoelectrogenic.

At the same time, despite the complex nature of interaction with microbial cells, the frequency of use of GR in microbial BFC is increasing [19]. However, from the data given in Ref. [38], one can see that even when using GR in BFC, their parameters are still inferior, and in some cases only approach the parameters of BFC based on enzymes.

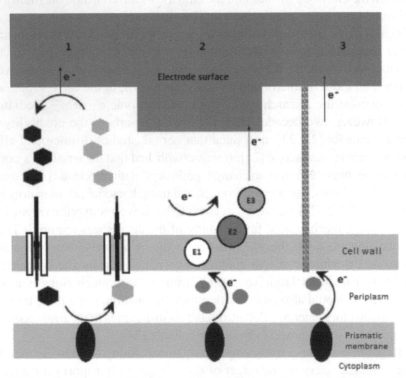

FIGURE 12.3 Schematic illustration of the mechanism of bacterial electron exchange using secreted mediators (1) (darker color – oxidized mediator, gray – reduced); surface cytochromes (2), designated as E1 – E3 and bacterial pilus (3) – nanowires or oxidoreductases.

12.3.3 GRAPHENE (GR) AS A MEMBRANE SENSOR

GR can be used as a sensitive element (membrane) of the sensor [30]. The NEMS sensitive element is a region of GR located above a cavity etched in a silicon oxide film on a silicon substrate. GR also lies on the electrical contacts, and the contacts, in turn, are combined into a Wheatstone bridge. If there is a pressure differential between the environment and the area inside the etched cavity, then the GR membrane deviates from the equilibrium position, and thus its state becomes strained. This leads to a change in the resistance of the GR region above the etched cavity, which leads to an imbalance of the Wheatstone bridge. Thus, by measuring the output voltage of the bridge, it is possible to determine the pressure acting on the membrane. Compared to traditional pressure sensors, pressure sensors, whose membranes are GR, demonstrate a greater sensitivity by several orders of magnitude, and also have much smaller overall and mass characteristics.

12.3.4 GRAPHENE (GR) AS A SUPERCONDUCTOR

Due to the two-dimensional structure and the possibility of chemical modification, GR is promising for the study of two-dimensional superconductivity. However, the superconducting state in single-layer GR has not yet been discovered, despite the existence of the effect in carbon-related carbon materials: intercalated fullerenes [31], graphite [33, 34] and carbon nanotubes [34].

Currently, an active search is underway for systems based on GR, in which superconductivity could be observed at temperatures above 1 K. The Li/ GR /Co_2Si/CoxSi/W system can be considered as a candidate for detecting superconductivity in doped single-layer GR. Intercalation leads to the formation of a solid solution of silicon in cobalt, as well as to the formation of the surface crystalline phase of cobalt silicide with the Co_2Si stoichiometry.

The interaction of GR with surface silicide is rather weak, which leads to a significant weakening of the hybridization of the electronic states of GR and cobalt and the appearance of the Dirac cone of states of GR near the Fermi level. This creates favorable conditions for studying the process of electronic doping of GR. Doping with lithium leads to a significant charge transfer – 0.165 electrons per unit cell of GR, which corresponds to an electron concentration of 3.1×1014 cm^{-2} [37]. This value is much higher than that achieved when lithium is doped with GR on the surface of gold, which demonstrates a significant influence of the choice of substrate material on the charge transfer in the system. In addition, the doping level is noticeably higher than for the GR/Ca/Au system, in which the estimate of the transition temperature to superconductivity gives a value of 1.5 K [35]. It is shown that the special form of the Fermi surface of the studied system creates favorable conditions for the resonant amplification of the electron-phonon interaction and the increase in the critical temperature.

12.3.5 USE OF GRAPHENE (GR) IN CAPACITORS

There are also developments of lithium-ion supercapacitors with a negative electrode based on carbon material. The possibility of obtaining three-dimensional porous structures allows us to consider GR as a promising active material of the anode of lithium-ion supercapacitors. Due to the porous structure and the large specific surface area of GR, its application can provide faster ion transport and a larger area where electrochemical reactions take place compared to graphite. Another advantage is the large specific capacity contributing to an increase in the specific energy of lithium-ion supercapacitors. The disadvantages of using GR in the manufacture of a negative electrode include a strong dependence of the potential on the degree of lithiation, which leads to a more dramatic change in the voltage of the lithium-ion supercapacitors in the process. Nevertheless, according to the Ragon diagrams given in [36], lithium-ion supercapacitors with a negative electrode based on GR show a large specific energy at equal power compared to lithium-ion supercapacitors with a negative electrode based on graphite.

12.4 CONCLUSIONS

GR is a versatile tool for improving the properties of various processes. The most promising, relatively inexpensive, and affordable method for producing GR of sufficiently high quality is chemical vapor deposition (CVD) on transition metals. Due to its properties, GR can be used in various industries to produce the most efficient devices and technologies.

In our opinion, the acceleration of the use of GR and materials requires substantial development of methods for mathematical modeling of the processes of formation of GR and its properties. Promising for this are the methods of multi-level modeling and application of machine learning methods. GR has particular prospects in the formation of new nano-structured materials, including metamaterials, the formation of which is possible both with a volumetric arrangement of GR inside the material and with a special distribution of GR over the surface of materials. These studies will be the task of further research by the authors.

ACKNOWLEDGMENTS

The works was carried out with financial support from the Research Program of the Ural Branch of the Russian Academy of Sciences (project 18-10-1-29) and budget financing on the topic "Experimental studies and multi-level mathematical modeling using the methods of quantum chemistry, molecular dynamics, mesodynamics, and continuum mechanics of the processes of formation of surface nanostructured elements and meta-materials based on them."

KEYWORDS

- **biofuel elements**
- **CVD method**
- **epitaxial cultivation**
- **graphene**
- **lithium-ion supercapacitors**

REFERENCES

1. Novoselov, S., Geim, A. K., Morozov, S. V., Jiang, D., Zhang, Y., Dubonos, S. V., Grigorieva, I. V., & Firsov, A. A., (2004). Electric Field Effect in Atomically Thin Carbon Films. *Science, 306*, 666.

2. Berger, C., Song, Z. M., Li, X. B., Wu, X. S., Brown, N., Naud, C., et al., (2006). Electronic Confinement and Coherence in Patterned Epitaxial Graphene. *Science, 312*, 1191–1196.

3. Reina, A., Jia, X. T., Ho, J., Nezich, D., Son, H. B., Bulovic, V., Dresselhaus, M. S., Kong, J., (2009). Large area, few-layer graphene films on arbitrary substrates by chemical vapor deposition. *Nano Lett., 9*, 30.

4. Chen, Z., Ren, W., Liu, B., Gao, S. P., Wu, Z. S., Zhao, J., & Cheng, H. M., (2010). Bulk growth of mono- to few-layer graphene on nickel particles by chemical vapor deposition from methane. *Carbon, 48*, 3543–3550.

5. Wu, Z. S., Ren, W., Gao, L., Liu, B., Jiang, C., & Cheng, H. M., (2009). Synthesis of graphene sheets with high electrical conductivity and good thermal stability by hydrogen arc discharge exfoliation. *Carbon, 47*, 493.

6. Fu, W., Kiggans, J., Overbury, S. H., Schwartz, V., & Liang, C., (2011). Low-temperature exfoliation of multilayer-graphene material from $FeCl_3$ and CH_3NO_2 co-intercalated graphite compound. *Chem. Commun., 47*, 5265.

7. Berger, C., Song, Z. M., Li, T. B., Li, X. B., Ogbazghi, A. Y., Feng, R., Dai, Z. T., Marchenkov, A. N., Conrad, E. H., First, P. N., & De Heer, W. A., (2004). Ultrathin Epitaxial Graphite: 2D Electron Gas Properties and a Route toward Graphene-based Nanoelectronics. *J. Phys. Chem. B, 108*, 19912.

8. Viculis, L. M., Mack, J. J., Mayer, O. M., Hahn, H. T., & Kaner, R. B., (2005). Intercalation and exfoliation routes to graphite nanoplatelets. *J. Mater. Chem., 15*, 974.

9. Potts, J. R., Dreyer, D. R., & Bielawski, C. W. R. S., (2011). Graphene-based polymer nanocomposites. *Ruof, Polymer, 52*, 5.

10. Banerjee, B. C., Hirt, T. J., & Walker, P. L., (1961). Pyrolytic carbon formation from carbon suboxide. *Nature, 192*, 450–451.

11. Karu, A. E., & Beer, M. J., (1966). Pyrolytic formation of highly crystalline graphite films. *J. Appl. Phys., 37*, 2179.

12. Mattevi, C., Kim, H., & Chhowalla, M., (2011). A review of chemical vapor deposition of graphene on copper. *J. Mater. Chem., 21*, 3324–3334.

13. Robertson, S. D., (1969). Graphite formation from low-temperature pyrolysis of methane over some transition metal surfaces. *Nature, 221*(5185), 1044–1046.

14. Bonanni, A., Loo A. H., & Pumera, M., (2012). Graphene for impedimetric biosensing. *TrAC-Trends of Analytical Chemistry, 37*, 12–21.

15. Dreyer, D. R., Park, S., Bielawski, C. W., & Ruoff, R. S., (2010). The chemistry of graphene oxide. *Chem. Soc. Rev., 39*, 228–240.

16. Liu, Y., Dong, X., & Chen, P., (2012). Biological and chemical sensors based on graphene materials. *Chem. Soc. Rev., 41*, 2283–2307.

17. Wu, H., Wang, J., Kang, X., Wang, C., Wang, D., Liu, J., Aksay I. A., & Lin, Y., (2009). Glucose biosensor based on immobilization of glucose oxdase in platinum nanoparticles/grapheme/chitosan nanocomposite film. *Talanta, 80*, 403–406.

18. Liu, C., Alwarappan, S., Chen, Z., Kong, X., & Li, C. Z., (2010). Membraneless enzymatic biofuel cells based on graphene nanosheets. *Biosens. Bioelectron, 25*(70), 1829–1833.

19. Shan, D., Zhang, J., Xue, H. G., Ding, S. N., & Cosnier, S., (2010). Colloidal laponite nanoparticles: Extended application in direct electrochemistry of glucose oxidase and reagentless glucose biosensing. *Biosens. & Bioelectron., 25,* 1427–1433.

20. Prasad, K. P., Chen, Y., & Chen, P., (2014). Three-dimensional graphene – carbon nanotube hybrid for high-performance enzymatic biofuel cells. *ACS AP. Mater. Interfaces, 6,* 3387–3393.

21. Potter, M. C., (1911). Electrical effects accompanying the decomposition of organic compounds. *Proceedings of the Royal Society of London, Series, B., 84,* 260–276.

22. Chaudhuri, S. K., & Lovley, D. R., (2003). Electricity generation by direct oxidation of glucose in mediatorless microbial fuel cells. *Nat. Biotechnol., 21,* 1229–1232.

23. Kim, H. J., Park, H. S., Hyun, M. S., Chang, I. S., Kim, M., & Kim, B. H., (2002). A mediator-less microbial fuel cell using a metal-reducing bacterium, Shewanella putrefaciens. *Enzyme Microb. Technol., 30,* 145–152.

24. Schröder, U., (2011). Discover the possibilities: Microbial bioelectrochemical systems and the revival of a 100-year-old discovery. *J. Solid State Electrochem., 15,* 1481–1486.

25. Malvankar, N. S., & Lovley, D. R., (2014). Microbial nanowires for bioenergy applications. *Curr. Opin. Biotech., 27,* 88–95.

26. Logan, B. E., & Regan, J. M., (2006). Electricity-producing bacterial communities inmicrobialfuel cells. *Trends Microbiol., 14,* 512–518.

27. Lovley, D. R., (2011). Live wires: Direct extracellular electron exchange for bioenergy and the bioremediation of energy-related contamination. *Energy Environ. Sci., 4,* 4896–4906.

28. Richter, H., Nevin, K. P., Jia, H., Lowy, D. A., Lovley, D. R., & Tender, L. M., (2009). Cyclic voltammetry of biofilms of wild type and mutant Geobacter sulfurreducens on fuel cell anodes indicates possible roles of OmcB, OmcZ, type IV pili, and protons in extracellular electron transfer. *Energy Environ. Sci., 2,* 506–516.

29. Pant, D., Bogaert, G. V., Diels, L., & Vanbroekhoven, K., (2010). A review of thesubstrates used in microbial fuel cells (MFCs) for sustainable energy production. *Bioresource Technol., 101,* 1533–1543.

30. Smith, A. D., Niklaus, F., Paussa, A., et al., (2013). Lemme electromechanicalpiezoresistive sensing in suspended graphene membranes. *Nano Lett., 13*(7), 3237— 3242.

31. Kelty, S., Chen, C., & Lieber, C., (1991). The superconducting energy gap of Rb3C60. *Nature, 352,* 223.

32. Hannay, N., Geballe, T., Mattias, B., Andres, K., & Schmidt, P., D., (1965). Superconductivity in Graphitic Compounds. *Macnair. Phys. Rev. Lett., 14,* 225.

33. Gruneis, A., Attaccalite, C., Rubio, A., Vyalikh, D. V., Molodtsov, S. L., Fink, J., Follath, R., Eberhardt, W., Buchner, B., & Pichler, T., (2009). Electronic structure and electron-phonon coupling of doped graphene layers in KC8. *Phys. Rev. B, 79*(205), 106.

34. Tang, Z., Zhang, L., Wang, N., Zhang, X., Wen, G., Li, G., Wang, J., Chan, C., Sheng, P., (2001). Superconductivity in 4 angstrom single-walled carbon nanotubes. *Science, 292,* 2462.

35. Fedorov, N., Verbitskiy, D., Haberer, C., Struzzi, L., Petaccia, D., Usachov, O., Vilkov, D., Vyalikh, J., Fink, M., & Knupfer, B. B., Gruneis A. (2014). Observation of a universal donor-dependent vibrational mode in grapheme. *Nature Commun., 5,* 3257.

36. Sanchez-Barriga J., Varykhalov A., Scholz M., Rader O., Marchenko D., Rybkin A., Shikin A., Vescovo E. (2010). Chemical vapour deposition of graphene on Ni(111) and Co(0001) and intercalation with Au to study Dirac-cone formation and Rashba splitting. *Diamond Related Mater., 19,* 734.

37. Yu, D., Usachev, F. A. V., Yu, V. O., Erofeevskaya, A. V., Vopilov, A. S., Adamchuk, V. K., Vyalykh, D. V., (2015). Formation and doping of lithium by graphene on the surface of cobalt silicide. *Solid State Physics, 57,* 5.

38. Reshetilov, A. N., Kolesov, V. V., Gubin, S. P., & Alferov, V. A., (2014). Application of graphene in bioflue elements. *Electrochemical Energy, 14*(4), 173–186.

Index

A

Academic rigor, 78, 82, 85–87, 92, 93
Acetone, 51, 58–61, 161
Aerosol, 182–185, 191–193
Alginate, 126
Aliphaticity, 166, 175
Alkali metals, 64, 65, 67
Alkaline solutions, 171
Aluminophosphate, 117
Aluminum, 25, 117, 148, 151
Ammonia, 157
Amplitude, 22, 23
Anaerobic digestion, 171
Angstroms, 18
Anions, 64, 66, 73, 106, 116, 118, 156, 157
Anisotropy, 25, 26
Antibiotics, 185, 186
Argon species, 154
Aromatic rings, 205
Aromaticity, 166, 175
Arsenic groundwater, 76, 88, 90, 91
Artificial
  intelligence, 35, 36
  neural network, 38, 39
  neuron, 41, 43, 47, 48
Atomic
  absorption spectrometry (AAS), 152,
    153, 155, 175
    advantages, 153
    limitations, 153
  emission spectroscopy, 154
  environment, 38
  force microscopy (AFM), 23, 131, 133
    175, 176
  nucleus, 160
  structure, 26
Auxiliary agents, 174

B

Bacteria, 185, 190
  cells, 205
  etiology, 181, 185, 195
Becke's three-parameter Lee-Yang-Parr
  (B3LYP), 2, 3
Bentonite, 81
Bicarbonate, 156
Biocatalyst, 205
Biodegradability characterization, 168
  analytical techniques, 169
    chromatographic analysis, 170
    gravimetric analysis, 169
    microscopic analysis, 169
    morphological analysis, 169
    physical/thermal analysis, 170
    respirometric analysis, 171
    spectroscopic analysis, 170
  biodegradation measuring methods, 169
  enzymatic techniques, 173
  microbiological techniques, 172
    clear-zone technique, 172
    direct cell count technique, 172
    pour plate/streak plate technique, 172
    turbidity determination, 173
  molecular techniques, 173
Biodegradable polymers, 168
Biodegradation, 168–172, 175
Biofilm, 169
Biofuel elements (BFC), 199, 203–206, 209
Biogas, 78
Biogradability, 171
Biomaterials, 112
Biomedicine, 121, 122, 124, 128
Biopolymers, 126
Bioremediation, 82
Biosorption, 82